The Frontiers Collection

The books in this collection are devoted to challenging and open problems at the forefront of modern science and scholarship, including related philosophical debates. In contrast to typical research monographs, however, they strive to present their topics in a manner accessible also to scientifically literate non-specialists wishing to gain insight into the deeper implications and fascinating questions involved. Taken as a whole, the series reflects the need for a fundamental and interdisciplinary approach to modern science and research. Furthermore, it is intended to encourage active academics in all fields to ponder over important and perhaps controversial issues beyond their own speciality. Extending from quantum physics and relativity to entropy, consciousness, language and complex systems—the Frontiers Collection will inspire readers to push back the frontiers of their own knowledge.

Shahen Hacyan

The Mathematical Representation of Physical Reality

Springer

Shahen Hacyan
Instituto de Física
Universidad Nacional Autónoma de
México (UNAM)
Mexico City, Mexico

ISSN 1612-3018   ISSN 2197-6619 (electronic)
The Frontiers Collection
ISBN 978-3-031-21256-7   ISBN 978-3-031-21254-3 (eBook)
https://doi.org/10.1007/978-3-031-21254-3

This Springer imprint is published by the registered company Springer Nature Switzerland AG
The registered company address is: Gewerbestrasse 11, 6330 Cham, Switzerland

Ce charme ! Il prit âme et corps,
Et dispersa tous efforts.

Que comprendre à ma parole?
Il fait qu'il fuit et vole!

—Rimbaud

To Betina

Preface

Mathematics has been impressively successful in describing a wide variety of phenomena in Nature. This success seems to be natural, but it is not obvious why it should be so; on the contrary, it is a complete mystery to the point that Paul Dirac—who unified relativity with quantum mechanics—came to think that mathematics and reality were two parallel but somehow interconnected worlds.[1] In this regard, Einstein once declared that "the eternal mystery of the world is its comprehensibility"[2] and Wigner, an eminent mathematical physicist, pointed out the "unreasonable effectiveness of mathematics in the natural sciences".[3]

Mathematical abstractions have been a fundamental step in the development of science, but at the cost of losing the direct perception of Nature which was so dear and familiar to our ancestors. In ancient times, contact with Nature was more immediate and without the mediation of complex devices. Abstract concepts were seldom used. Paradoxically, the world became objective through abstraction, as science found its support in mathematical idealizations.

Nevertheless, the mathematization of science was not successful in all areas. The remarkable advancement of biology, for instance, has shown that a mathematical description does not necessarily bring about a fundamental result, as in the study of inanimate matter. Even in the realm of physics, many natural systems are too elusive to lend themselves to an exact quantitative description. Of course, the difficulties are even more evident in human sciences, where the inescapable intervention of the subject inhibits an objective and quantifiable interpretation. At the very most, statistical data can be accumulated and correlated between them, but without an underlying mathematical theory.

In any case, it has been possible to study and understand many physical processes with mathematical language: its equations, functions, variables, integrals, and the various parameters identified with space, time, mass, energy, field, etc. However,

---

[1] Dirac (1939).

[2] Einstein (1936).

[3] Wigner (1960).

though mathematical reality is an excellent representation of the world, it is not identical to it.

This book deals with the irruption of mathematics in physical sciences, mainly from Galileo and Newton to the present days. The text is divided into two parts. In the first one, the reader will find a brief history of how mathematics was introduced into physics—despite its "unreasonable effectiveness"—and the criticisms it received from earlier thinkers. The second part has a more philosophical approach, for it intends to shed some light on that mysterious effectiveness. For this purpose, I review the debate on the existence of innate ideas that allow us to understand the world and also the philosophically based opinions for and against mathematics. In this context, Schopenhauer's conceptions of causality and matter are very pertinent, and their validity will be revised in the light of modern physics. The final question is whether the effectiveness of mathematics could be explained by its "existence" in an independent platonic realm, as Gödel believed.

Mexico City, Mexico   Shahen Hacyan

Acknowledgment I wish to express my deepest gratitude to Beatriz Loria Lagarde for her invaluable assistance in the preparation of this text and many helpful comments and criticisms.

Contents

Part II

Part I

Chapter 1
Classical Period

Abstract The Pythagorians believed in mathematics as the key to Nature, but it is with Galileo and Kepler that its importance began to be realized. Nonetheless, Descartes had doubts about the utility of mathematics. In this respect, the refraction of light provoked a discussion between Fermat with his mathematical approach and the French Cartesians. Newton's *Principia* was a landmark in science, but it was not easily accepted since the motion of planets had "only" been described, but the mechanical origin of gravity had not been explained. The use of an analytical rather than a geometrical approach was an important contribution of Euler to the science of Mechanics.

1.1 In Illo Tempore

It is said that Pythagoras was the first man who ever perceived a close relationship between numbers and the physical world. According to legend, he discovered that musical notes were harmonious to the ear if the lengths of the strings of a lyre obeyed certain numerical relationships (2 to 1: octave; 3 to 2: fifth, etc.). It is also said that he wanted to apply this discovery to other fields of knowledge; thus, he searched for a relationship between the motion of the planets and the harmonic notes. His disciples said that he could hear the music of the heavenly spheres.

According to Arthur Koestler, in *The Sleepwalkers*,

> Nobody before the Pythagoreans had thought that mathematical relations held the secret of the universe. Twenty-five centuries later, Europe is still blessed and cursed with their heritage. To non-European civilizations, the idea that numbers are the key to both wisdom and power, seems never to have occurred.[1]

Thus, for Koestler, the great difference in scientific development between Eastern and Western civilizations is due to the importance given to mathematics, even though both civilizations may be proud of their achievements in other fields. At any rate, the fact is that the East has successfully adopted, in the last century, the mathematics born in the West, and even made important new contributions.

---

[1] Koestler (1959).

S. Hacyan, *The Mathematical Representation of Physical Reality*,
The Frontiers Collection, https://doi.org/10.1007/978-3-031-21254-3_1

As for ancient Greece, a century after Pythagoras, some of his teachings remained in the philosophers' memory. "No one enters without knowing geometry" is Plato's alleged inscription at the entrance to his Academy, but it is not clear what he conceived as geometry. According to his disciple Aristotle, Plato thought of geometry as a reality existing between the material world and the world of Ideas (we will return to this subject in the second part of this book). As for Aristotle himself, he had, within his vast work, a treatise on physics in which he explained all natural phenomena with arguments of the common sense, but without ever trying to quantify or relate them to geometry.

Euclid, who lived between the fourth and third centuries BC, was undoubtedly the greatest mathematician of antiquity. To him we owe the rigorous style of postulating a few axioms that seem obvious to almost everyone and, based on them, develop increasingly complex consequences following rigorous logical reasoning. Euclid's method was purely geometrical, for it was still very far from the development of algebra. Even in his treatment of numbers, such as prime numbers, he relied on geometric representations: for instance, how many times a line segment of a given length fits into another segment.

Following Pythagoras and Euclid, Archimedes (c. 287–c. 212, BC) from Syracuse (now in Sicily) contributed notably to the application of mathematics to the study of natural phenomena. He is credited with the invention of various machines, mainly war machines, but also his treatises on geometry, in which he taught how to calculate surfaces and volumes of geometric figures, are also famous. As for the applications to practical problems of physics, he correctly described the mechanism of the lever and taught how to locate the center of gravity of various bodies using Euclid's geometry.

We may wonder how ancient peoples managed, without a good numerical system, to express very large numbers. Archimedes wrote a short essay on the subject, *The Sand Counter*, in which he estimated the number of grains of sand that could fit into what, at his time, was believed to be the Universe. Though counting was quite complicated before the Arabic system, Archimedes used the concept of a *myriad*, which is equal to 100 times 100, that is, 10,000, and conceived a myriad of a myriad, which he called a myriad of a second. order, that is 10,000 times 10,000, or 10^8 in modern notation. With that procedure, he could conceive even larger numbers: myriads of third order, fourth order, etc., and estimate the volume of the Universe and the number of grains of sand it might contain. Since at that time it was believed that the universe was limited by the sphere of the fixed stars, whose radius was thought to be about 100 million stadia (the Greek stadium was equivalent to about 180 of our meters), Archimedes concluded that it would take something like a "sixteenth-order myriad" of grains of sand to fill the universe. A quantity that can only be imagined with the aid of numbers, and that nowadays seems very simple to conceive with the help of an exponential notation.

After Archimedes, other important developments in mathematics, with applications to the study of Nature, had to wait several centuries. Undoubtfully, the most important one was the invention of Algebra, which is due largely to Persian and Arab mathematicians, who also contributed to other disciplines, such as optics: the study of light. But Pythagoras's original project, to understand the world with numbers, was just beginning to take shape. It was not until the time of Galileo that the original Pythagorean idea that numbers are the key to the Universe was recalled.

1.2 Galileo

Galileo Galilei (1564–1642) assigned to mathematics a fundamental part in the study of Nature. One of his best-known popular books, *Il Saggiatore* (*The Assayer*), begins with the oft-quoted words:

> Philosophy is written in that great book, the Universe, which is continually open before our eyes... that book is written in the language of mathematics, and its characters are triangles, circles and other geometric figures without which it is humanly impossible to understand a word.

Galileo was one of the main proponents of a new world view based on logical reasoning and, above all, experimentation to confront a mathematical description with the real world.

In another of his major works, the *Dialogues on the Two Great World Systems*, Galileo presented three characters, one of whom, Simplicio, unsuccessfully defends Aristotle against Salviati, through whose mouth Galileo expounds his own ideas about Nature. Simplicio takes up Aristotle's criticism against Plato's supposed passion for mathematics, since "... mathematical subtleties work very well in the abstract, but not when applied to sensitive and physical matters." But Salviati-Galileo argues that the sensible world would be incomprehensible without a well-structured conceptual apparatus that allows us apprehend reality through thought. Furthermore, if we are to believe Plato, it would be a reality that is known before birth, or, as modern philosophers would say, an a priori knowledge, prior to experience (we will also return to the subject in the second part of this book). Geometry is a good example of that other reality. Geometrical objects—circles, straight lines, etc.—do not exist in a pure form in the sensible world, but the objects we perceive are more or less circular or straight thanks to the geometrical ideas we possess. Perhaps this is the reason why Galileo defended Copernicus' heliocentric theory so passionately, even though, in practice, it did not describe the motion of the planets more precisely than Ptolemy's model. But the important point was that the planet should move in circles around the Sun because the circle was the most perfect geometrical figure (even though it was necessary to move it slightly from the center of the Earth's orbit to fit the astronomical data).

Galileo argued that sensory experience could not be fully trusted. Although observations are essential to guide and test a theory, the senses can deceive and lead to erroneous conclusions. The best example is the motion of the Sun, which appears to revolve around the Earth, as it was believed for centuries and millennia.

Of course, Galileo carried out multiple experiments on Earth and was the first to study the heavens with a telescope, thus laying the foundations of a new science of nature. With his telescope, he discovered the satellites of Jupiter, the sunspots and the rotation of the Sun, the phases of Venus, the mountains and valleys of the Moon, and the wealth of stars that make up the Milky Way. He made it very clear that a theory cannot follow just from an accumulation of experimental data, but also from a mathematical structure that permits those data to be ordered, thus transcending sensory experience.

However, the mathematics known and used by Galileo were very rudimentary compared to what was developed later. In the absence of a well-established measurement system, his calculations were restricted to simple arithmetic operations for calculating the proportions between one process and another. For instance, the lapse of time for a body to fall from a certain height compared to the lapse of time to fall from another height. Galileo would have been greatly impressed, had he seen the further mathematical achievements in the description of physical processes.

It is also well known that Galileo had serious problems with the Church for stating that the Sun does not turn around the Earth, but the other way around, just as the satellites of Jupiter turn around that planet. It is said that he invited important religious leaders to look through his telescope to verify, with their own eyes, the veracity of his discoveries, but these personalities were not convinced by what they saw. However, that anecdote must be put in the context of his time. Paul Feyerabend, in *Against Method*,[2] pointed out that Nature, in ancient times, was not studied with artificial means, since what could not be perceived directly with the senses was mistrusted. Nowadays, we accept without the slightest doubt that there are things invisible to the naked eye, such as atoms, microbes, galaxies… but in Galileo's time it was not at all obvious that an instrument did not create illusions. Ultimately, according to Feyerabend, the Church was defending a world view that ordinary men could easily understand, without the intervention of sophisticated instruments.

Although the telescope made it possible to increase the size of objects seen on Earth, it was not obvious that the images of celestial objects, which had never been seen before, could correspond to something real. If Galileo thought he was observing new stars where nothing could be seen with the naked eye, there was no way to corroborate their existence. Furthermore, there was still no good theory of how a telescope works. Galileo had made one, but he had succeeded through trial and error. It would not be until 1611, shortly after his observations, that his colleague Kepler published his *Dioptrice*, in which he graphically explained the functioning of a telescope. Besides, Galileo's telescopes were quite primitive, so a certain amount of imagination was necessary to see what he claimed to see. Undoubtedly, he had the enormous merit of correctly imagining much of what he reported, but not everybody was able to see with his telescope everything he claimed to see.

In any case, a new era began with Galileo, in which the five common senses were no longer sufficient to correctly perceive the world. It was necessary to resort to artificial means and mathematical calculations… that only experts knew how to

---

[2] Feyerabend (1975).

handle. The new way of studying the world turned out to be extremely successful, but many thinkers, without denying its validity, preferred a more subjective view of the world. We will return to the subject in Chaps. 10 and 11.

1.3 Kepler

Despite its novelty, Galileo's promising vision regarding the usefulness of mathematics was not shared by all his contemporaries... and with some reason, as we have just seen. An example of the wrong path that can be taken by relying too much on mathematical ideas was Kepler's attempt to explain the planetary system, as in his early work, the *Mysterium Cosmographicum*. Kepler based his speculations on the fact that there are five regular solids, called platonic, which are three-dimensional bodies, with their surfaces formed by flat regular polyhedrons. These regular solids— tetrahedron, cube, octahedron, dodecahedron, and icosahedron—have the property of fitting perfectly inside a sphere and fitting another sphere inside. From this geometrical property, Kepler speculated that, if there are five regular solids, it would not be a coincidence that there were also five planets (those known at the time: from Mercury to Saturn). Thus, he claimed that the orbits of the five planets should correspond to the five spheres that fit to each platonic solid, one inside the other, with the Sun in the center according to Copernicus' heliocentric model. He thus thought to have unveiled the mystery of the creation of the cosmos through geometry!

Of course, the celestial harmony that Kepler believed to have discovered turned out to be a mere illusion. He himself realized it and decided to take the problem of planetary motion more seriously. Thus, through an incredibly patient and precise analysis of the observations that his teacher Tycho Brahe had inherited him, Kepler discovered the famous three laws that bear his name. These laws would later give Newton the key to discover universal gravitation and thus claim mathematics as the language of Nature.

In an essay he wrote on Kepler,[3] the celebrated physicist Wolfgang Pauli (who we will meet later) argued that the great astronomer was a representative (one of the last?) of the medieval worldview. He was looking for "a magical-symbolical description of nature", in comparison with the more rational visions of Galileo and Newton, based on the study of Nature and the discoveries of its laws. But Pauli's main argument is that Kepler, unconsciously, used innate ideas expressed through "archetypes", those mental structures typical of a "collective unconscious" according to the psychoanalytical theories of Carl Jung (with whom Pauli had a close friendship). Celestial harmony, built by God himself, would be Kepler's attempt to harmonize the world with his own mental structures. To understand Nature would be through "a 'matching' of inner images pre-existing in the human psyche with external objects and their behavior".[4] Moreover, said Pauli, geometry would be for Kepler an archetype of the

---

[3] Pauli (1949).

[4] *Ibid.*

world, a belief that led him to accept astrology as justified by the harmony between Earth and the heavens, a harmony that could be discovered by geometrical means.

We will return to the subject of innate ideas and mental structures in Chap. 11. Let us note for moment that all magical-symbolic interpretations of nature were gradually superseded by a description of Nature in mathematical terms. We can consider Kepler as the intermediary between the medieval view of nature and the new science promoted by Galileo.

1.4 Light

In addition to the motion of the planets, the nature of light was also the subject of study—and much speculation—by scholars of all ages. In ancient Greece, there were two theories in this regard: according to one, vision was due to a kind of "visual fire" emanating from the eyes, which would "feel" objects, as a blind man feels something with his walking stick. According to the other theory, light would have its source outside and would penetrate the eyes after bouncing off the objects.

Euclid himself was a supporter of visual fire and, based on the geometry he knew so well, wrote an extensive treatise on optics to explain the visual appearance of objects—such as the apparent distance-size relationship—supposing that rays emanate from the eyes and propagate in a straight line. As a problem in geometry, Euclid's treatment was unobjectionable since it contained the theoretical bases of the laws of perspective. Later, the great astronomer Ptolemy wrote another treatise on optics with the same ideas as Euclid, in which he also applied geometry to various optical problems. These treatises, taken as exercises in geometry, are perfectly valid since it is only necessary to change the direction of the propagation of the rays: from the object to the eyes and not the other way around.

To the Arab philosopher and mathematician Ibn Al-Haytham, also known as Alhazen (965–1040 approx.), is due a comprehensive seven-volume treatise on optics in which, among many other things, he refuted the theory of light emitted by the eyes. Alhazen was a pioneer of the science of optics. He also used geometrical methods to study the reflection of light off a curved surface, and also the way it enters the eyes and what happens inside them.[5]

Refraction is a particular phenomenon of light that consists of the deviation of a ray when penetrating a transparent medium. In Galileo's time, an empirical formula for the angle of deflection of a ray passing from one medium to another was already known. It is now called Snell's law, attributed to the Dutch mathematician Willebrord Snell (1580–1626), although it was discovered much earlier by Ibn-Sahl, a predecessor of Alhazen. According to this law, the sine of the angle of incidence is always proportional to the sine of the angle of refraction.[6] It is worth noticing that

---

[5] For a very detailed and complete history of optics, see Darrigol (2012).

[6] Snell rediscovered this law in 1621, but never published it. It was Huygens who mentioned it in his treatise.

the explanation of this law gave rise to lengthy discussions, quite independently of mathematical methods. It was Christiaan Huygens (1629–1695) who gave a mathematically rigorous explanation of the phenomenon in his famous *Traité de la lumière (Treatise on Light)* of 1678. He proved with perfect mathematical rigor that Snell's law can be deduced from the principle that light is a wave, a wave that propagates somewhat slower in a transparent medium that is denser than air. His proof was based on an impeccable geometrical construction: the assumption that light propagates as a spherical wavefront, which breaks as its speed changes. But, of course, one had to accept that light is a wave, which was still a hypothesis at that time.

1.5 Descartes

To René Descartes (1596–1650) we owe a new version of philosophy, quite different from the narrow scholastic dogmas that still prevailed in his time. In 1637, he published his famous *Discours de la méthode (A Discourses on the Method)*, with which he wanted to break away from all philosophical prejudices and start the study of nature from the very beginning and from simple and clear concepts. To this purpose, he recommended, as an example worth following, the method used by geometers:

> The long chains of reasonings, every one simple and easy, which geometers habitually employ to reach their most difficult proofs had given me cause to suppose that all those things which fall within the domain of human understanding follow on from each other in the same way... [...] considering that of all those who had up to now sought truth in the sphere of human knowledge, only mathematicians have been able to discover any proofs...[7]

However, mathematics for Descartes was just an example of how the study of other sciences should be approached with a rigorous method, even though they did not have any practical use! This was clearly expressed in the same paragraph quoted above: "Though I would not expect any other utility than to accustom my mind to being satisfied with truths and not settling at all with false reasons."

The *Discourses of the Method* were the introduction to the treatises on *Dioptrics, Meteors* and *Geometry*. Although Descartes was a great mathematician, the creator of analytic geometry—expounded in the third of these treatises—, he applied mathematics only marginally to the study of Nature. It is enough to peruse through the other two treatises to verify that, apart from the inevitable figures, no mathematical equation or formula appears (unlike what is found in a modern book on optics or meteorology). In the *Meteors*, he correctly explained the formation of the rainbow, but only qualitatively, based on an ingenious experiment performed with a glass ball imitating the drops of water in the air, which are at the origin of the rainbow. He also discovered the decomposition of the solar light into the colors of the rainbow many decades before Newton, but he did not draw any quantitative conclusion from his finding.

---

[7] *Discours*, Second Part, (translation by Ian Maclean).

In the *Dioptrics*—one of the first optical studies in Europe, prior to that of Huygens—Descartes deduced Snell's law from a geometric figure, without explicitly mentioning the trigonometric relationship between the angles of incidence and reference. The explanation he gave of this phenomenon was not at all rigorous, as it was based solely on an analogy with the trajectory of a bullet that decreases its speed when penetrating a denser medium from a less dense one, such as, for instance, from air to water.

1.6   About Refraction

Descartes' explanation of the phenomenon of refraction did not convince (and for good reasons) his contemporaries, particularly the English philosopher Thomas Hobbes (1588–1679)—author of the famous treatise on political philosophy *Leviathan*—, who had also dabbled in natural philosophy and had his own theory of optics. At that time, most scholars used to communicate with each other through their correspondence with the Jesuit friar Marin Mersenne (1588–1648), established in Paris. Hobbes, in a letter sent to Mersenne but aimed to Descartes,[8] strongly criticized the treatment of the French philosopher, pointing out how absurd it would be to suppose that a bullet (or a ray of light), upon reaching the surface of water, decided to change its direction... as if it had a will of its own!

This critical comment by Hobbes displeased Descartes, who wrote to Mersenne in response:

> [Hobbes] uses a very frivolous reasoning to refute what I wrote... since, he says, it would follow that a bullet has a knowledge of the laws of geometry. As if, because something happens in nature according to the laws of geometry, it follows that bodies obey those laws by having understanding or knowledge of them.[9]

However, the strongest objection against Descartes's hypothesis was to come from his compatriot and contemporary Pierre Fermat (1601–1665), a distinguished mathematician, co-inventor with Descartes of the analytic geometry (and a notary by profession). Fermat proposed a very different treatment of the problem of refraction that, in the long run, not only turned out to be correct, but also laid the foundations for what would become a fundamental principle of mathematical physics. However, Fermat could not avoid receiving criticisms in the same terms as Hobbes.

1.7   The Principle of Least Action

Against Descartes' supposed demonstration of Snell's law, Fermat proposed another one, based on the principle that "nature always acts by the shortest path." Suppose, he said, that a ray of light goes from point M to point N (see Fig. 1.1); it is well

---

[8] Hobbes to Mersenne (1641).

[9] Descartes to Mersenne (1641).

Fig. 1.1 Refraction of light:
from M to N

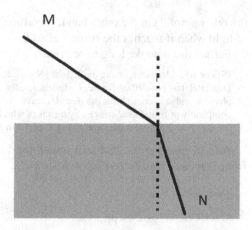

known that it moves in a straight line in the same medium, since the straight line is the shortest line. Now, what happens if point N is inside a transparent medium, like water or glass, and point M is outside? Experience shows that light does not follow a straight path. Then, Fermat postulated, the real path must be the one that the light travels in the shortest possible time.

To support his claim, Fermat proposed to solve the following problem: a bullet travels from point M, in a certain medium, to point N in another medium, its speed being greater in the first medium than in the second. What is the path it would take to get from M to N in the shortest possible time?[10] Using very ingenious geometrical arguments (differential calculus had not been invented), Fermat deduced the correct formula for the deviation of the ray starting from his initial hypothesis.[11] He published his result in a memoir of 1662,[12] and communicated it by letter to some of the distinguished intellectuals of the time. As expected, the Cartesians did not like his solution to the problem, since Fermat had assumed that the speed of light is lower in a denser medium, just the opposite of what Descartes had postulated for a bullet in his loose analogy. Thus, a distinguished Cartesian, Monsieur Clerselier,[13] wrote to Fermat[14] to inform him that his principle of least time was "a moral principle and by no means a physical one," an argument reminiscent of the Hobbes-Descartes dispute. Indeed, how can a ray of light, once it reaches water, decided in which direction to head to get to a certain point as soon as possible? One would have to suppose that the ray remembers that it started from such a point "with the order to search, upon meeting another medium, for the path that it could travel in the shortest time to reach

---

[10] It is the same problem of the lifeguard and the drowning swimmer. The lifeguard (from M) runs faster on the beach and swims slower in the water, so he must follow a path as in the figure to reach the swimmer (at N) as quickly as possible.

[11] For details of the deduction, see Dugas (1950), Chap. V.

[12] *Synthesis and refractions,* cited by Dugas, *op.cit.*

[13] Claude Clerselier (1614–1684) was the editor and translator of several works of Descartes.

[14] Clerselier to Fermat, May 6, 1662.

the other point." On the other hand, according to Descartes, a force deflects the ray
of light when it reaches the surface of water.

Fermat did not take long to respond

[S]ince you, Monsieur, assure me that she [Nature…] is content with the course that Monsieur
Descartes has prescribed for he, I wholeheartedly abandon you my presumed conquest of
physics, and it is enough for me that you leave me in possession of a problem of geometry
completely pure and *in abstracio*, by means of which one can find the path of a body passing
through two different media and trying to perform its motion as quickly as possible.[15]

And in case Fermat had said something false, he ended his letter with a quote
from Torcuato Tasso's *Jerusalem liberated*:

Quando sará il vero

Si bello, che si possa a ti preporre?

(When will the truth be

So beautiful, that it can be preposed to you?)

As we see in this historical anecdote, the idea that mathematics could be used to
understand the world was not universally accepted, despite the fact that, as Galileo
had stated, it was the language of nature. At that time, natural philosophers preferred
a description of the world in terms of simple processes drawn from everyday experiences; with mechanical examples and analogies, what we could nowadays call
"mechanical philosophy".

1.8 Mechanical Philosophy

During the seventeenth and eighteenth centuries, natural philosophers intended to
explain the most basic physical phenomena by means of elementary mechanical
principles, such as the indestructibility and impenetrability of matter, the collision
between the particles that compose it and, particularly, the assumption that the interaction between bodies could only be given through something material in between.
This interpretation of natural laws can be characterized as mechanical philosophy:
an attempt to explain natural phenomena with mechanical models based on physical
processes that are understandable, or seem to be so, because they are part of everyday
experience. The explanation in mathematical terms would be complementary, but it
would not be expected to reveal anything new or more profound.

Following a mechanical conception, Descartes conjectured that the world is filled
with a more or less subtle matter, made up of minute particles of various sizes, in
constant motion, colliding with bodies and pushing them. To explain the motion of
the planets, he assumed that the finer elements of that matter would form gigantic
eddies around the Sun that would drag the planets into their orbits, with smaller

---

[15] Fermat to Clerselier, May 21, 1662. Also cited in Dugas, *op. cit.*

eddies around the planets that would move their satellites. All of which was not hard to imagine, since the image of a whirlpool of water dragging tree leaves or corks in circles is quite familiar; it seemed natural to generalize this picture to the planets.[16] In this regard, it is noteworthy that Descartes made no quantitative attempt to describe planetary motions by means of a mathematical formalism. The laws discovered by his contemporary Kepler were irrelevant to his world system and he didn't even bother mentioning them anywhere.[17]

1.9 Kepler (2)

Another example of a mechanical explanation, without resorting to mathematical formulations, is the attempt of the great scientist Huygens to explain gravity. At a session of the Royal Society in London around 1690, he presented an experimental model which he believed would explain gravity as the result of the motion of a subtle matter à la Descartes. The experiment, described in detail in his *Discours de la cause de la pesanteur* (*Discourse on the Cause of Gravity*) published in French in 1690, consisted of submerging pieces of wax in a vessel filled with water and observing that, as the liquid was stirred, the wax fragments plunged to the bottom. Huygens did not try to explain it quantitatively, but he speculated that something like that might be the cause of gravity.[18]

In summary, a characteristic of mechanical philosophy is to dispense with a mathematical description as much as possible and look for a qualitative description based on processes that seem natural. Natural because they are part of our everyday experience. Later, we will see that there are still residues of mechanical philosophy in modern physics.

1.10 Vis Viva

A few months before Newton sent his manuscript of the *Principia* to the printer, his illustrious contemporary Gottfried Wilhem Leibniz (1646–1716) published a memoir in the *Acta Eruditorum* with a striking title: *A Brief Demonstration of a*

---

[16] R. Descartes, *Le monde ou traité de la lumière* (*The world or treatise on light*), Chaps. VII–XI. Written between 1632 and 1633 and published posthumously in 1662.

[17] Kepler published his first two laws in 1604 and his third one in 1619. *Harmonices Mundi* appeared in 1619.

[18] Perhaps Huygens came up with this idea when he was having tea with his English hosts, since he could have noticed that stirring his drink with a spoon caused the tea leaves to plunge to the bottom of the cup. What happens, actually, is that, due to the friction with the cup, the centrifugal force is weaker near the walls and the fluid is pushed towards the center. But, of course, that has nothing to do with gravitation.

Notable Error of Descartes and Others Concerning a Natural Law,[19] in which he criticized the concept of "quantity of motion", which the Cartesians identified with the product of mass and speed. It was not only a semantic question, since it was about defining a "motive force" that obeys a law of conservation in nature, since "if God conserves the same quantity of motion in the world", it is necessary to specify what we are talking about.

Leibniz asked the question: what is preserved in the rise or fall of a body? We must start from the fundamental postulate that the "force" required to lift a body of one pound to a height of four feet is the same as that required to lift a body of four pounds to a height of one foot. But the speed with which a body falls, according to Galileo, is proportional to the square root of the distance traveled.[20] Therefore, if the "force" to lift a body is conserved, it must be proportional to the product of the mass m and the square of the velocity v. Consequently, for the "force" (mass multiplied by height) to be the same for any mass and height, what should be constant in all possible cases is

$$mv^2.$$

Some years later, Leibniz would call it *vis viva*. The reader will recognize it as twice what we now call kinetic energy. The concept of *vis viva*, or "living force," turned out to be very useful as the definition of a quantity that is naturally conserved in all physical processes. Actually, Leibniz's concept of "force" corresponds to what is currently known as energy, while, as we shall see later, Newton's "force" remained the force as understood in contemporary physics.

1.11   Newton

Mathematics, and very particularly geometry, burst into the natural sciences in a spectacular way, in 1687, with the fundamental treaty of Isaac Newton (1642–1727): *Philosophiæ Naturalis Principia Mathematica*. A new style of science was initiated that would definitively prevail in the future. The method followed by Euclid, based on definitions, axioms, and theorems, became an indisputable paradigm for the description of physical processes.

Newton first specified what should be the concept of force, from where he formulated the basic laws that define it—his famous three laws that every student must learn in school. The first law is a definition of what force is not: a body moves in a straight line and conserves its momentum if no force is applied to it. And what is momentum? Newton defines it precisely as the product of mass and velocity. The

---

[19] *Acta Eruditorum*, March 1686. English translation in: *G. W. Leibniz, Philosophical Papers and Letters* (1956).

[20] More precisely, $v^2 = 2gh$, where g is the acceleration due to gravity and h is the initial height. The *vis viva* turns out to be proportional to the height h.

second law, which can be interpreted as a definition of force, states that the change in the momentum of a body is just the force applied to it. Finally, the third law postulates that to every applied force corresponds a force of reaction (for instance, the Sun attracts the Earth and the Earth attracts the Sun, although this latter effect is hardly perceptible due to the difference in mass of the two bodies). Additionally, Newton postulated the parallelogram law for the composition of forces, which is fundamental to any problem in two or three dimensions.

Thus, having clarified what is meant by force and how it produces motion, Newton demonstrated with mathematical rigor that the observed orbits of the planets are due to a universal law of gravitation. With the mathematical formalism of his invention, the calculus of fluxions (predecessor of the differential and integral calculus by Leibniz), he was able to derive Kepler's three laws in the First Book of the *Principia*. It was the final blow to the Cartesian theory of vortices, although they still survived for some time in the fatherland of their author.

As for the concept of mass, it turned out to be more complicated than it might seem at first glance. We will return to it later, but for now let us note that Newton, in the *Principia*, defined mass as "quantity of matter," which is "the measure of [matter] itself, arising from its density and volume taken together." It seems, then, that density was for Newton a more primary concept than mass. Nowadays, various definitions of mass can be found in physics textbooks, for instance as the "amount of matter", although it is never specified how such an amount can be measured. Moreover, there is often a general confusion between mass and weight. We know that weight, although related to mass, is the manifestation of the gravitational force exerted by the Earth and disappears into outer space. Obviously, this was unknown in ancient times, and size was even confused with weight, since large bodies tend to weigh more than small ones (Jammer 1961). Galileo understood that mass is the inertia that a body exerts against the attempt to move it and he noted, very correctly, that this inertial mass is always proportional to weight, which is why all bodies fall with the same acceleration, as can be deduced from his famous (imaginary) experiment on the tower of Pisa. Newton was aware of the equivalence between inertial mass and weight, and he even confirmed it experimentally with pendulums of equal length but different weights and materials; he verified that they all oscillated with the same period. It is now known as the *equivalence principle*, which was essential for Einstein to develop his theory of gravitation, as we will see in the following chapters.

After publishing the *Principia*, Newton lost interest in natural philosophy, being more occupied with matters of theology and history according to the Bible.[21] He waited until 1704, when his hated colleague and critic Robert Hooke (1635–1703) died, to publish his next great work: *Opticks*, an important treatise that would be a reference point for all studies of light. In this book, Newton did not use his characteristic mathematical treatment of the *Principia* but devoted most of the text to a qualitative description of experiments with light. Not a single mathematical formula is found throughout the text, but only figures that describe his experiments.

---

[21] For a very complete biography of Newton, see Westfall (1980).

It is a curious fact that dealing with the refraction of light, Newton erroneously stated (like Descartes) that the speed of light is higher (!) in a transparent medium than in a vacuum. He managed to explain—but wrongly!—Snell's law (he did not derive it explicitly but left it as an elementary exercise in geometry), with the assumption that the component of the velocity perpendicular to the surface of the medium on which it falls must be higher.[22]

1.12 Reactions

Despite his indisputable achievements, Newton's work was not received by scholars of his day with the enthusiasm that might be expected. To begin with, it must be remembered that there was a rivalry between English and continental philosophers at the time of the publication of the *Principia Mathematica* (Newton must have named it so to distinguish his work from Descartes's *Principia Philosophiae*), the French being particularly fervent devotees of Descartes. Serious criticism came from scholars such a Huygens and Leibniz, who acknowledged the merits of the theoretical synthesis achieved by their English colleague but insisted that the cause of gravitation was far from being elucidated. A description with mathematical formulas is not an explanation in terms of known and understandable concepts. The real problem for Newton's contemporaries was: how can two material bodies attract each other through apparently empty space without the mediation of something substantial? For to move a body, it is always necessary for another material body (a rope, the wind, etc.) to have contact with it; any other cause of motion would be magic or witchcraft. However, celestial bodies attract each other without anything visible or palpable between them. Even if Descartes' theory of vortices was wrong, at least it tried to explain why planets move without resorting to some mysterious action through an empty space. With his mathematical calculations, Newton had only shown *how* heavenly bodies move.

Newton himself was clear about the problem and emphasized in his defense that, in fact, he never pretended to know the ultimate reason for the gravitational attraction. He had only tried to describe its effects mathematically. The problem of the origin of gravity was left to others. "I do not build hypotheses" (*Hypothese no fingo*) is his famous phrase that he included in the General Scholium of the second edition (1713) of the *Principia*. Still, Newton believed that something like a subtle matter should produce gravity, as he stated in his treatise on optics, published in 1704, and in his correspondence with Bentley from 1692 to 1693 in which he discussed the cause of gravity and action at a distance.

In summary, physics was consolidated as a theoretical science thanks to the *Principia Mathematica*. However, this text was far from being comprehensible to everyone. It is said, quite rightly, that in Newton's time only a few initiates were able to fully understand it… and nowadays no one would be advised to study mechanics

---

[22] *Opticks*, Book Two, Part III. Proposition 10.

with it.[23] Even so, the basic concepts that appeared in the *Principia* were profound enough to survive to this day, albeit in a very different mathematical language. Leibniz must be credited for a different version of the calculus of fluxions in algebraic and not geometrical terms, a formalism that was perfected by the Bernoullis[24] and gave rise to what is now known as differential and integral calculus. Notable mathematicians, such as the Bernoullis themselves and Leonard Euler expanded Newton's work to further advances in mathematical mechanics.

1.13 Euler

To Leonhard Euler (1707–1783) we owe most of the applied mathematics used today, together with many fundamental contributions to pure mathematics.[25] In fact, there is hardly any branch of mathematics to which he did not contribute, and the theoretical mechanics pioneered by Newton is not an exception.[26]

Among numerous contributions, Euler elucidated the concept of "mathematical function"[27] (he invented the notation $f(x)$ that appears in basic texts), with which he could express trigonometric concepts—such as sines, cosines and tangents—, in the form of functions, defined in terms of infinite series. Likewise, he interpreted *logarithms* as functions and established numerous unsuspected properties, thus defining the fundamental number e (2.71828 ...) as the basis of natural logarithms, a number which, together with the famous π (*pi*, the notation is also his), appears conspicuously in many mathematical formulas.

Euler also dealt with what is (misleadingly) called an "imaginary number"— square root of a negative number—, whose interest and usefulness were under discussion at his time. Defining the imaginary number i (in modern notation[28]) as the square root of *minus one*, $\sqrt{-1}$, he proved that the concept of a real number can be consistently generalized to that of a complex number, which is the sum of a real number and an imaginary number. Thus, Euler discovered a vast mathematical continent, which can be condensed in his famous and surprising formula,

$$e^{i\pi} + 1 = 0,$$

---

[23] S. Chandrasekhar (1995) wrote *Newton's Principia for the Common Reader* (his last book) with the noble intention of rewriting Newton's geometrical proofs in a form understandable to the "common reader" (the question remains whether he succeeded).

[24] Mainly the Jacob brothers (1654–1705) and Johann (1667–1748) Bernoulli, and the next generation: the Nicolaus I cousin (1687–1759) and Daniel (1700–1782).

[25] Dunham (1999), Calinger (2019).

[26] Suisky (2009).

[27] The rigorous definition of the concept of function, to the liking of mathematicians, had to wait until the nineteenth century. See Youschkevitch (1976).

[28] Engineers prefer to use the letter j and reserve i to the electric current.

that relates the most basic numbers among them: e, π, 0, 1, and the imaginary i. Furthermore, he found an unexpected relationship between geometry and algebra, by showing that trigonometric functions (sinus and cosinus) can be expressed in terms of complex functions of complex numbers by means of an important formula relating trigonometric functions to exponents and imaginary numbers[29]:

$$e^{ix} = \cos x + i \sin x,$$

which is a generalization of the previous formula. As we shall see later, complex functions will be essential in the formulation of quantum mechanics in the twentieth century.

The science of mechanics also owes much to Euler. As already mentioned, Newton's original approach had been purely geometric: all his proofs were based on a cumbersome fluxion calculus that required quite a bit of ingenuity to follow. Euler found a more comfortable way to solve problems in mechanics: he reformulated Newtonian physics with his own theory of functions, combined with differential and integral calculus in the version of Leibniz and the Bernoullis, and thus freed mechanics from the narrow framework of geometry. He laid the foundations of what would be known as analytic mechanics, where differential and integral calculus, and particularly partial differential equations,[30] replaced geometric theorems. It was in 1736 that Euler published, in Latin, his important treatise on mechanics, *Mechanica sive motus scientia analytice exposita*, in which he applied the new science to various cases of practical utility, not considered by Newton, such as the motion of extended solid bodies.

With respect to Newton's second law, Euler argued in his treatise that the primary concept should be force and not mass, since the mass of a body is measured from its motion produced by a force of a given magnitude. Consequently, Euler postulated the mass as the ratio between the force and the acceleration it produces. However, it is not obvious that force is a primary concept and that it is possible to measure mass without reference to force. Also, regarding the second law, Euler clearly pointed out in his treatise that the motion of a body is necessarily due to some reason, and that this reason is nothing other than what we call force. See his Proposition 7 in the *Mechanica*:

A body must remain in a state of absolute rest unless it is acted on by an external cause to move it.[31]

---

[29] An earlier version of this formula, in terms of logarithms and not exponentials, is due to Roger Cotes (1682–1716), a disciple of Newton.

[30] The partial derivative of a function of two or more variables is the derivative with respect to only one of those variables, holding the others constant. For instance, if $f(x, y)$ is a function of x and y, the partial derivative with respect to x is written $\partial f/\partial x$ in modern notation.

[31] *Corpus absolute quiescens perpetuo in quiete perseverare deben, nisi a causa externa ad motum sollicitetur.*

Euler also made important contributions to hydrodynamics. He published a treatise in 1757 in which he deduced the basic equations that describe the flow of a fluid.[32] In a sense, he foreshadowed the concept of "field", since the fluid is described by its density, pressure, and velocity as a set of functions that depend on both time and position in space. This is not the case with a point particle or a solid body in classical mechanics, whose position in space is just a function of time, which is the only physical variable. Partial differential equations appear in this treatise,[33] since it is necessary to describe the properties of the fluid as they vary not only in time but also in space.

As for the already mentioned mechanistic attempt to explain all natural phenomena, it is worth mentioning Euler's speculations on the origin of forces in nature. According to him, all forces were due to collisions between solid particles, as if they were billiard balls. In an essay written in 1750 to defend this idea, he stated: "There are no more forces in the world than those that have their origin in the impenetrability of bodies."[34] In that, he was following the dominant ideas of his time.

---

[32] *Principes généraux du mouvement des fluides*, Opera Omnia: Series 2, Vol. 12, pp. 54–91.

[33] See Footnote 27.

[34] *Recherche sur l'origine des forces*. Mémoires de l'Académie des Sciences de Berlin, 6, 419–447 (1750).

Chapter 2
Enlightenment

Tu m'appelles à toi, vaste et puissant génie,
Minerve de la France, immortelle Émilie.
Voltaire

Abstract Newton's work was highly appreciated by the French philosophers and mathematicians of the period called Enlightenment, particularly through the treatises of D'Alembert and Lagrange. Hamilton formulated what is now known as Analytical Mechanics and the important Principle of Least Action. The appearance of the crucial concept of energy in the middle of the nineteenth century relegated force to a secondary role.

Shortly before Newton died, the great writer, poet, and philosopher Voltaire (1694–1778) took temporary refuge in England to escape the climate of intolerance at homeland. There he witnessed the funeral of the famous English natural philosopher and was impressed that a scientist in England received such a funeral worthy of kings.[1] Although he was not a mathematician, Voltaire realized the importance of Newton's work and decided to expand it to France. For this purpose, he resorted to his scientific friends and compatriots, such as D'Alembert, Maupertuis, Alexis Clairaut and, specially, Émilie Marquise du Châtelet.[2]

Previously, in 1740, the Marquise du Châtelet had published a long popular treatise, entitled *Institutions de Physique*, in which she discussed in an accessible form (the text was dedicated to her son) the basic ideas of the new science of mechanics: the concepts of space and time, the nature of matter, the motion of bodies, gravity and Newton's contributions. The treatise ended with a discussion on the "living forces" and the "dead forces". The former being the forces that produce motion, while the latter are due only to constraints and are cancelled by reaction forces according to Newton's third law (for instance, a body sliding on an inclined surface is subject to

---

[1] Voltaire, *Lettres philosophiques*.

[2] Jean le Rond D'Alembert (1617–1783), Pierre Maupertuis (1698–1759), Alexis Clairaut (1713–1765) and Émilie de Breteuil, Marquise du Châtelet (1706–1749).

the force of gravity and, in addition, to a "dead force" compensated by a reaction that prevents it from penetrating the surface).

The Marquise, who was very close to Voltaire, undertook the great task of translating Newton's *Principia* from Latin into French, to make that great work more accessible to her countrymen. Her translation was published posthumously in 1759 and is, to date, the only French translation.[3] Voltaire wrote a preface ("This translation that the most learned men of France should have made... a woman undertook and completed it to the astonishment and glory of her country..."), and Émilie herself wrote an appendix in collaboration with Clairaut, another French advocate of Newton.[4] The appendix included a long Commentary (*Commentaire*) in which they summarized the contribution of the English scientist and solved various problems of physics, such as the shape of rotating bodies, the origin of the tides and the refraction of light. The French version of the Marquise was the source of inspiration for French mathematicians in the second half of the eighteenth century.

2.1 D'Alembert

In his highly influential *Traité de Dynamique*, published in 1743 (and republished in 1758), D'Alembert began by noting:

> Bodies act on one another in three different ways: i) by collision, ii) by the intermediacy of some other body that joins them, iii) by means of a reciprocal attraction, such as gravity.

"Newton took much care of this third class of action" wrote D'*Alem*bert, hence he would restrict himself to the study of bodies that collide or are constrained to move in some way. A more terrestrial mechanics, one could say, which was to be the beginning of mechanics applied to practical problems on Earth and not only in the heavens.

For D'Alembert, the following principles were the important ones to be taken as basis: (i) the principle of inertia (Newton's first law), (ii) the law of the composition of forces or law of the parallelogram; (iii) the laws of equilibrium between bodies (action and reaction: essentially Newton's third law). The parallelogram law is of utmost importance, as D'Alembert noted, for it permits to deduce the total force resulting from two (or more) forces, a principle to which Newton devoted only a corollary, just after stating his three laws.

Through the principle that now bears his name, D'Alembert showed how to separate the so-called "living forces" from the "dead forces", following the arguments of the Marquise du Châtelet. Of course, dead forces are irrelevant for the motion of the planets as Newton studied them, but it is essential to take them into account for practical problems of mechanics and engineering on Earth. For this purpose,

---

[3] Émilie died in 1749 of puerperal fever, a few days after giving birth. She finished the translation during what was to be her last pregnancy.

[4] Clairaut accompanied Maupertuis on his expedition to Lapland (see below).

D'Alembert relied on the concept of "virtual displacement", which is the displacement that a body is restricted to realize—at least in principle—according to the constraints imposed on it. D'Alembert solved the problem of the "death forces" using the parallelogram theorem to separate them from the "living forces", just noticing that the former are always perpendicular to the direction of the motion, that is, to the virtual displacements.

D'Alembert's treatise will pave the way for the mathematically more elaborate formalism of Lagrange.

2.2 Maupertuis and the Least Action

Among the "French Newtonians" we must also mention Maupertuis, who became renowned as the "man who flattened the Earth" after his expedition to Laponia, where he measured the Earth 's meridian and confirmed that our planet is flattened at the poles, just as Newton had predicted.[5]

As a mathematician, Maupertuis formulated in a new way the principle of least action that Fermat had visualized. In a memoire addressed to the *Académie des Sciences* in 1744,[6] he took up the Cartesians' criticism of Fermat's deduction, since he was not convinced that light was slower in a denser medium. Fermat would have been wrong, but Maupertuis graciously conceded that something could be rescued from his hypothesis that Nature looks for "the simplest paths",[7] and he was willing to correct Fermat in his deduction of the Snell's law. Thus, Maupertuis calculated the trajectory of light passing from one medium to another according to his own definition of action. He obtained the correct result, but with the hypothesis that the speed of light is greater in a denser medium. Actually, it all depends on the definition of action: according to Fermat, what should be minimum was the travel time, which is the distance *divided* by the speed, but the action, according to Maupertuis, was the distance *multiplied* by the speed. As a result, Maupertuis obtained the same formula as Snell, but with the speeds of light reversed (in any case, the speed of light in a medium was completely unknown in the eighteenth century... it was necessary to wait for Foucault[8]).

In a following article,[9] Maupertuis applied his principle of least action to the collision between perfectly hard or elastic particles. He noted that the triple product

---

[5] Voltaire dedicated a poem to him: «Vous avez confirmé dans des lieux plein d'ennuis/ce que Newton savait sans sortir de chez lui...» (You have confirmed in places plenty of nuisances /what Newton knew without leaving his home...).

[6] Dugas, p. 250.

[7] Maupertuis, *Accord de différentes loix de la nature qui avoient jusqu'ici paru incompatibles* (1744).

[8] The speed of light in water was measured by Léon Foucault (1819–1868) in the 1850s.

[9] Maupertuis, *Les loix du mouvement et du repos déduites d'un principe metaphysique* (1746).

mass × *velocity* × *distance* of his original action is the same as *mass* × *velocity-squared* × *time*, so the action cán also be defined as Leibniz's *vis viva* multiplied by the time taken by the motion. Next, Maupertuis showed that the actual change in speed is the smallest possible, and thus he obtained the correct speed of a body after the collision. For him, the correct way to express this law would be that the "amount of action" is minimal, with the action defined as the triple product of mass, speed, and distance. This would be the quantity that God commands to be minimal in all physical processes, which would be, according to Maupertuis, the proof of the existence of a Supreme Being.[10] In modern terms, Maupertuis' action is the integral of the momentum mv along a physical path, $\int mv \, ds$, which is the same as the integral over time $\int m \, v^2 \, dt$ (since the differential of the distance $ds = v \, dt$).

2.3 Lagrange

As we have already remarked, the mathematical formalism of the *Principia*, based on Euclid's geometry, is not very practical since not everyone possesses the genius and geometric intuition of Newton. It was therefore necessary to find a more routine method to attack the problems of mechanics. The formalism that was finally imposed, and that physicists and engineers use to date, is due to the scientists of the eighteenth century mentioned above, and particularly to Joseph-Louis Lagrange (1736–1813), to whom belongs the honor of having completely mathematized mechanics in a purely analytical way, without resorting at all to geometrical figures.

Originally from Turin, Italy, Lagrange[11] arrived in Paris as a distinguished member of the Académie des Sciences. There he published his masterpiece, *Mécanique Analytique*, which appeared in French a year before the storming of the Bastille, and in which he proposed a new formalism based exclusively on algebraic analysis. In the preface (*Avertissement*) he wrote[12]:

> There are already several treatises on mechanics, but the plan of the present one is totally new. I proposed myself to reduce the theory of this science, and the art of solving related problems, to general formulas, the simple development of which gives all the necessary equations for the solution of each problem.

Thus, its author gave a transcendental and definitive turn to Newtonian physics, since

> No figures are to be found in this work. The methods that I expose requires neither constructions, nor geometrical or mechanical reasoning, but only algebraic operations, subject to a

---

[10] Voltaire ridiculed Maupertuis' theological-scientific speculations in *Diatribe du docteur Akakia*, which he wrote during his stay in Berlin. The publication of the pamphlet disgusted his host, King Frederick II of Prussia, who held Maupertuis in high esteem and had invited him to preside over the Berlin Academy of Sciences. Voltaire was forced to flee back to France.

[11] Originally Giuseppe Lodovico Lagrangia.

[12] My translation.

regular and uniform process. Those who like Analysis will gladly see Mechanics turned into a new branch and will be grateful to me for thus extending its domain.

Hence, from the very preface, he warned his readers that he would not use any figure and, effectively, he kept his promise: not a single figure appears throughout the five hundred pages of this treatise. The breaking away from the geometric formalism of the *Principia*, with its abundance of geometric figures, was radical, but Lagrange's formalism turned out to be far superior and is the one that is still used to this day (but do not try to follow his original notation, which is appalling in comparison with the present-day formulation that relies on the compact vectorial notation,[13] developed at the end of the nineteenth century which greatly simplified calculations[14]). In any case, it was the definitive farewell to the geometric constructions that were beyond the reach of most mortals, and it was the beginning of the mechanization of calculation. It was enough to state a problem correctly to automatically obtain the differential equations which solved it! Mechanics had been reduced to mathematical manipulations.

Newton's second law—force equals mass times acceleration—is, in practice, a set of second order derivatives of the position with respect of time (i.e., the acceleration $\frac{d^2r}{dt^2}$, where r is the position vector and t the time). In three dimensional space, if the position of a particle is $r = (x, y, z)$, in Cartesian coordinates, then

$$m\frac{d^2x}{dt^2} = F_x, \quad m\frac{d^2y}{dt^2} = F_y, \quad m\frac{d^2z}{dt^2} = F_z,$$

where m is the mass of the particle and F_x, F_y, F_z are the component of the force in each of the three directions in space. Given the force $F = (F_x, F_y, F_z)$, "all" we have to do is solve the above equations. In Lagrange's formalism, Newton's second law takes an apparently more complicated but totally general form, since the constraints imposed on a physical system are taken into account from the very beginning thanks to the D'Alembert principle. Additionally, one can choose the most appropriate system of coordinates. Thus, for instance, if one wants to study the motion of a particle constrained to move on the surface of a sphere (say, the Earth), it is more convenient to use longitude and latitude as coordinates, instead of simple Cartesian coordinates.

Lagrange showed that the famous *vis viva* (with an appropriate factor 1/2) plays a fundamental role, together with another mathematical function that he designated with the letter Π, which is related, through its first spatial derivatives, to the forces that produce a given motion. Such a function would be identified some years later as the *potential energy*, a fundamental concept that came to replace that of force. The heart of the Lagrangian formalism is a mathematical function of all the positions and velocities of the particle under study, in *any coordinates* system, the so-called *Lagrangian function L*, which is essentially the difference between (1/2) the *vis viva*

---

[13] A vector V can be represented as an arrow. In three dimensions, $V = (V_1, V_2, V_3)$, where V_i ($i = 1, 2, 3$) are the components of the arrow along the three basic directions of the space.

[14] See, e.g., Crowe (1967) for the history of the vector analysis.

and the potential energy.[15] Explicitly, if we have N "degrees of freedom", that is, we need N coordinates q_i ($i = 1, 2 \ldots N$) and no more to describe a physical system, the Lagrangian is as function

$$L(q_i, v_i, t)$$

of the coordinates q_i and the velocities $v_i = \mathrm{d}q_i/\mathrm{d}t$, which are the time derivatives of q_i.[16] If one manages to construct this mathematical function from the characteristic property of a physical system to be studied (a particle, a solid body, a fluid…), one can automatically deduce all the relevant equations of motion—the Lagrange equations which generalize Newton's equations—and thus solve any problem of mechanics … in a mechanical way.

As for the principle of least action, Lagrange obtained it directly from his equations. This principle, that Maupertuis seemed to have pulled out of his sleeve, was fully justified by Lagrange, who obtained it as a simple corollary from the laws of mechanics.

Lagrange's most distinguished disciple was Pierre Simon Laplace (1749–1827), who also made important contributions to the science of mechanics. In his monumental treatise on the World System (*Exposition du système du monde*), Laplace studied the motion of the planets in the Solar System, considering not only the gravitational attraction of the Sun, but also the mutual attractions between the planets. Laplace succeeded to show that the Solar System is stable under small perturbations, thus correcting Newton who believed that the Universe required a Divine intervention from time to time to rectify the slight deviations in the orbits of the planets (a concept that his rival Leibniz harshly criticized…[17] but that's another story.)[18]

2.4 Hamilton

After Lagrange, William R. Hamilton (1805–1865) made a further breakthrough in the science of analytical mechanics. Based on Lagrange's formalism, Hamilton developed a new mathematical approach very suitable to solve a wide variety of problems (including quantum mechanics, as we will see later). Instead of considering the

---

[15] In modern terms, it applies to a "conservative force", which can be expressed as the vectorial gradient of a potential function. A conservative force is such that the work performed in moving a particle from A to B does not depend on the trajectory but only on the initial and final positions A and B. Gravity is a conservative force.

[16] And also time t in some cases.

[17] "God Almighty would want to wind up his watch from time to time… He would not have, it seems, sufficient foresight to make it a perpetual motion… The machine of God's making, so imperfect, that he is obliged to clean it now and then by an extraordinary concourse, even to mend it, as clockmaker mends his work…" *Leibniz-Clarke Correspondence*.

[18] According to an oft-cited anecdote, Napoleon once asked Laplace why God is not mentioned anywhere in his treatise, to which Laplace answered: "Sir, I did not require that hypothesis.".

velocity of an object, he defined a generalization of the momentum which turned out to be most convenient in obtaining the equations of motion. Hamilton's momentum was no longer Newton's "quantity of motion" (mass times velocity), but something more general, including the position in space.[19] His article had the suggestive title: *On a General Method in Dynamics; by which the Study of the Motions of all free Systems of attracting or repelling Points is reduced to the Search and Differentiation of one central Relation, or characteristic Function.* In Hamilton's formalism, the central role is played by the mathematical function we now know as the *Hamiltonian*, which is essentially equivalent to the concept of energy but broadly more general: it is a function of the position and Hamilton's "generalized momenta". Of course, both formulations, Lagrange's and Hamilton's, are equivalent and the use of one or the other depends on the problem to be solved. Lagrange's equations of motion, like Newton's original ones, are second-order differential equations for the position in space[20]; Hamilton's equations are of first order, at the cost of doubling the number of equations: one set of equations for position and another set for generalized momenta. In modern terminology, the lagrangiano is basically (1/2) the *vis viva minus* the potential energy, while the Hamiltonian is (1/2) the *vis viva plus* the potential energy and it can be identified in most cases with the total conserved energy. The former is a function of position and velocity, and the latter is a function of position and generalized momentum.

As for the principle of least action, Hamilton showed convincedly that it is a fundamental law of Nature. In a paper published in 1833 in Dublin, Hamilton studied the problem of geometrical optics and derived Fermat's principle in a completely general form. Then, being on track, he extended this principle to material bodies and proved that the equations of motion of mechanics follow rigorously from the condition of minimal action. The precise definition of the action must be the integral of the Lagrangian taken along *any* trajectory given by $q_i = q_i(t)$ (the coordinates as functions of time):

$$\int L(q_i, v_i, t) dt.$$

It turns out that this action is minimum for those trajectories that are the real physical ones, which are precisely the unique trajectories that satisfy Lagrange's equations. This principle is known today as the "Hamilton principle" for good reasons.

Thus, the principle of least action ceased to be the subject of metaphysical discussions and became a respectable member of the physical sciences. The formalism also revealed a clear mathematical relationship between mechanics and optics through its basic equations, which turned out to be similar despite describing different physical processes. Furthermore, Hamilton's formulation was supplemented by Carl Jacobi

---

[19] Hamilton defined the generalized momentum as the partial derivative (see note 33) of the Lagrangian L with respect to the velocities: $p_i = \partial L / \partial v_i$.

[20] That is, they contain second derivatives with respect to time.

(1804–1851), to give rise to a new formalism in which the action is calculated directly from a differential equation. The Hamilton–Jacobi formalism is currently considered the culmination of analytical mechanics, as it is the most general way of solving problems in classical mechanics. The motion of bodies has thus been reduced to a purely mathematical problem.

If Archimedes claimed that he could move the Earth had he a lever and a fulcrum, present day theoretical physicists may well claim that they can solve any problem of the physical world if they have the correct Lagrangian or Hamiltonian.

2.5 Force or Energy?

With the new formulations of Lagrange and Hamilton, the force, which Newton had defined as that which changes the quantity of motion, disappeared to make way for what is currently known as energy. Newton did not mention anything like that in all his work, and although the word "energy" has always been in common use, it was without the precise meaning of modern physics. In fact, this fundamental concept appeared in physical theories in the mid-nineteenth century, when physicists appropriated that word to designate a new concept, motivated mostly by the steam engines that were being developed at that time. Energy in physics is a mathematically defined quantity that has the important property of remaining constant in many processes. By identifying this mathematical entity with the name energy, physicists were able to speak colloquially of the "conservation of energy"… and they realized that there should be some conserved quantities in the physical world, in addition to the quantity of matter.

In 1841, Julius Robert Mayer (1814–1878) published a seminal article entitled *On the Quantitative and Qualitative Determination of Forces*, in which he first mentioned the idea that in all physical processes there must be a quantity that is conserved. In that article, Mayer used the word force (*Kraft*) to designate what would later be known as energy. A few years later, in 1847, Hermann von Helmholtz (1821–1894) published an important treatise on the *Conservation of Force*, where he used the same word, force (*Kraft*), in the same sense as Mayer. Helmholtz argued that "force" should be conserved in all processes and only change its form. His proposal attracted the attention of William Thomson, later known as Lord Kelvin, who developed his own theory of what he renamed *energy*, from the Greek ενέργεια. Shortly thereafter, his colleague and compatriot Walter Rankine (1820–1872) further specified the term by defining "real energy" (actual energy) as (1/2) Leibniz's *vis viva* and "potential energy" as the function Π defined by Lagrange without giving it a name. As explained by Rankine himself, he was inspired by Aristotle's ideas about *dynamos* and *energy*. Lastly, in 1867, Thomson and Tait's *Treatise of Natural Philosophy* appeared—a fundamental book for the physicists of their time—, in which "real energy" changed its name to "kinetic energy", as it has been definitively accepted.

Thus, a mathematical concept, with a well-chosen name, permitted to see from a new perspective a whole series of phenomena (motion, heat, work done by a machine, etc.) that until then had appeared in disconnected ways. With the elucidation of such a fruitful concept, a new branch of physics appeared, thermodynamics, concerned with the use of heat to perform mechanical work. Sadi Carnot (1796–1832) described the relation between heat and work, with the steam engines in mind,[21] and Rudolf Clausius (1822–1888) developed the mathematical formulation of what Carnot had qualitatively described in his essay. Thus, in analogy with energy, Clausius defined a new concept, *entropy*, a word derived from the Greek, εργον (action, work) and τροπή (change, alteration). Entropy can be interpreted as a measure of the energy that performs work and that can no longer be used. Hence, Clausius was able to formulate, in the form of equations, the basic laws of thermodynamics that relate quantities such as energy, work, temperature, heat… and entropy to each other. The two basic laws of thermodynamics can be summarized as: (1) The total energy of the Universe is constant. (2) The entropy of the Universe always increases. The latter makes it possible to define, at least in theory, a direction of time.

As for the concept of force, which was so fundamental for both Newton and Euler, it practically disappeared due to the lack of an incontrovertible definition of force and mass. While acceleration is an intuitively clear concept since it is a combination of space and time, it is not clear whether the primary concept should be force or mass. Should force be defined as the product of mass and acceleration ($F = ma$), according to Newton, or should mass be defined as the ratio of force to acceleration ($m = F/a$), according to Euler?

The problem is that force is measured in terms of mass, and vice versa, in a kind of circular definition.

In any case, the concept of force began to lose its importance in the second half of the nineteenth century, due to the appearance of energy (and the field, as we will see later). So much so that Gustav Kirchhoff (1824–1887), in his *Treatise on Natural Philosophy*,[22] remarked that the particular motion of a body depends on its initial position and velocity, while the acceleration, which is the change in velocity,[23] has a general form for all motions and does not depend on the initial conditions. In other words, Newton's second law is just another way of stipulating that acceleration—change in velocity—has a universal character, independent of the initial position and velocity that determine each particular motion. For Kirchhoff, force was just a convenient mathematical concept.

Ernst Mach (1838–1916) also referred to the concept and history of force in his famous treatise *The Science of Mechanics*.[24] He pointed out that, before Galileo, force was related to pressure (a stone is moved by the pressure exerted on it), and that Galileo's great contribution was to relate it rather to the acceleration. This allowed Newton to give a precise mathematical definition of what was meant by force. Mach

---

[21] S. Carnot, *Réflexions sur la puissance motrice du feu et sur les machines propres à développer cette puissance*.

[22] *Vorlesungen über mathematische Physik, Mechanik.*, (Leipzig: BG Teubner, 1876). Cited by Dugas, *op. cit.* Chap. X, §4.

[23] The second derivative of position with respect to time.

[24] Mach (1883).

was also critical of force being a primary concept since it was indissolubly tied to mass. He even devised a scheme for deducing the relative masses of two interacting bodies from their accelerations,[25] but the method is not practical at all.

Henri Poincaré (1854–1912),[26] in an essay on the foundations of classical mechanics,[27] pointed out that "the idea of force is a primitive, irreducible, indefinable notion; we all know what it is, we have the direct intuition of it. This intuition comes from the notion of effort, which is familiar from childhood." Likewise, he noted that a definition of force, to be of any use, "must explain us how to measure it." Poincaré then discussed various ways of measuring a force or, equivalently, a mass, but he was unable to reach any conclusion and had to admit finally the failure of any attempt to define mass. Therefore, he proposed a definition "which is nothing more than a confession of impotence: the masses are coefficients that are convenient to introduce in the calculations."

Summing up, as Jammers (1961) remarked, due to the new positivist attitude at the end of the XIX century: "What once, in Newtonian physics, played a central role, was now considered an obscure metaphysical notion to be banished from science."

In any case, it is fair to recognize that the concept of force gave rise to other basic concepts, such as the principle of least action, associated with Fermat and Maupertuis; Leibniz's living force; D'Alembert's principle; Lagrange's analytic mechanics; Hamilton's principle, the Hamilton–Jacobi formalism, etc.; up to the concepts of energy and field, that emerged in the mid-nineteenth century. Force was replaced first by energy, then (as we shall see in the following) by the field, and finally it disappeared from relativity and quantum mechanics, the two fundamental theories of the twentieth century. Instead, what turned out to be two unavoidable concepts, which have survived the effects of time, are the principle of least action, whose importance was not suspected at the time, and the more recent concept of energy. It is important to note that this former principle and the concept of energy can only be rigorously defined by means of mathematics. Without mathematics, they would be only vague and abstract concepts expressed in common language.

---

[25] Mach, *po. cit*, Chap. II, § V.

[26] We will return to Poincaré in the chapter on modern physics.

[27] H. Poincaré, *La mécanique classique*, in *La science et l'hypothèse*, Chap. VI (Flammarion 1968).

Chapter 3
Electricity and Magnetism

The partial differential equation entered theoretical physics as a handmaid, but has gradually become mistress.
Einstein (1949), p. 63.

Abstract The relation between electricity and magnetism was realized by several scientists in the beginning of the nineteenth century, most particularly by Faraday who discovered many of the laws of these phenomena through his crucial experiments. The formulation of electromagnetism in mathematical form is due to Maxwell, who synthesized all its known laws in partial differential equations. Faraday and Maxwell introduced the fundamental concept of field.

With the advances in physics in the nineteenth century, the old discussions about hidden qualities of matter, action at a distance, and the nature of the forces began to wane, but the mechanism that produces gravity continued to challenge natural philosophers, who could not believe in an interaction between distant bodies without the intermediacy of "something".[1] As a result, the mechanistic vision continued to be valid for some time, coexisting with an increasingly complex mathematical formalism. However, the pretension to explain gravity as a purely mechanical effect was gradually abandoned when the interest of natural philosophers shifted towards electrical and magnetic phenomena, in which a mysterious action at a distance was also clearly manifested, but with the enormous advantage that, unlike gravity, it was possible to experiment with it in many possible ways. Even in Newton's time, the phenomena of electricity and magnetism attracted little attention from natural philosophers, but the situation began to change at the beginning of the nineteenth century.

---

[1] For a history of the development of the action-at-a-distance concept, see, e.g., Hesse (1962), Hacyan (2004).

3.1 Amber and Magnets

The ancient Greeks knew that if a piece of amber (*elektron*) is rubbed with a cloth, it attracts bits of straw. They also knew that some stones found in a certain region called Magnesia had the curious property of attracting iron. Petrus Peregrinus, who lived in the thirteenth century, wrote a brief treatise in which he reported his experiments with magnetic stones; he noticed, particularly, that a magnet has two different poles, which repel or attract each other depending on whether they are, respectively, equal or different.[2] Three centuries later, in 1600, the first detailed study of magnetism and electricity appeared: *De magnete*,[3] by the English physician and natural philosopher William Gilbert (1544–1603). In his treatise, Gilbert studied magnetic forces and deduced, quite correctly, that the Earth is a gigantic magnet. He also experimented with the force of attraction produced by the friction of amber (the word *electric* is due to him) and concluded that it was different from a magnetic force.

Gilbert described the results of his measurements in detail, but without proposing a physical explanation for the phenomenon of magnetism. It was Descartes, always keen to explain all natural processes, who speculated that magnetism could be due to certain minute particles ridged with hooks that penetrated the pores of the iron and grabbed its particles.[4] These ridged particles would be like screws that turn in one direction or another, thus producing the attraction or repulsion of the poles of a magnet. Of course, this description was only qualitative, although Descartes can be credited with having experimented with iron filings, like Peregrinus, and observed what would be known much later as "lines of force".

In any case, in Gilbert's and Descartes' days, electricity and magnetism were mere curiosities. With the sole exception of the compass, they seemed to have no other use than performing sleights of hand. It was in the age of the Enlightenment that scientists began seriously to experiment with electric and magnetic forces. For this purpose, the invention of the Leyden jar, predecessor of the condenser, was of great help, since it permitted to store electrical charges in a controlled way, instead of rubbing pieces of amber. Another important instrument was the voltaic pile, invented by Alessandro Volta (1745–1827), with which electrical currents could be generated. In its original version, the voltaic battery consisted of a succession of copper and zinc plates, separated by wet cardboards, which produced an electric current by chemical reactions. Finally, natural philosophers began to realize that gravity is not the only fundamental force in the inventory of Nature, and that the electrical and magnetic phenomena were worth studying.

---

[2] *Epistle of Petrus Peregrinus: De Magnete.* Philosophical Research Bulletin, Vol. 6, No. 7 (2017), pp. 277–307. https://archive.org/details/letterofpetrusp00pieriala/page/n5/mode/2up.

[3] *De Magnete, Magnetisque Corporoibus, et de Magno Magnete Tellure: Physiologia noua, Plurimis & Argumentis, & Experimentis Demonstrata.* «Of the magnet, of the magnetized bodies, and of that great magnet the Earth: A new physiology demonstrated with many arguments and experiments».

[4] *Principes de Philosophie, Livre IV*. Published in French in 1647.

In the eighteenth century,[5] it had already been noticed that there are two kinds of electricity, which were originally called *vitreous* and *resinous*, because one was produced by rubbing glass and the other by rubbing amber. They repelled each other if they were of the same stuff (glass-glass or amber-amber) and attracted each other if they were different (glass-amber). It was the big difference with the gravitational force, which is always attractive. To that respect, Charles-Augustin de Coulomb (1736–1806)[6] performed the first experimental verification of the electric force between two charges and verified that it also obeys an inverse square law, like gravity. Coulomb used a torsion balance for his experiment, an instrument that can measure very small forces with great precision. It seems, however, that Henry Cavendish (1731–1810), among other distinguished Englishmen, had already verified this law, but they did not receive sufficient attention from their colleagues on the Continent. A curious fact is that the experiment was repeated in modern times with an instrument like the one used by Coulomb, but without reaching a convincing result.[7] Probably, Coulomb expected an inverse square law, analogous to Newton's law, and that is what he saw according to his intuition... which turned out to be correct.

An important step in the eighteenth century was the realization of electricity having important dynamical effects, and not restricted to attraction or repulsion only. There had to be something equivalent to an "electrical fluid": a flow of electricity. The best example was lightning, which is a gigantic electrical discharge in the atmosphere, as verified by Benjamin Franklin (1706–1790), who thus invented the lightning rod. Franklin also correctly deduced that there are not two kinds of electricity but only one, manifested by its excess or deficit; he called them positive and negative, which sounds more scientific than vitreous and resinous.[8]

As for the electrical fluid, Luigi Galvani (1737–1798), in Italy, performed a famous experiment to reveal that electricity has the curious property of moving the muscles of a dead frog (which surely inspired the novel of Dr. Frankenstein). At first, Galvani thought that there was a kind of "animal electricity" in the body of animals, a conjecture that motivated some arduous discussions, particularly with his compatriot Alessandro Volta. According to Volta, there was an electrical fluid of external origin and, to prove it, he built the first battery that produces electrical fluid in a controlled way. Thanks to the voltaic pile, scientists were able to study in more detail the phenomena associated with electricity and magnetism.

---

[5] Since our main topic is not the history of physics in general, but the applications of mathematics, we will only mention the most important contributions to the study of electricity and magnetism. For a detailed history up to the nineteenth century, see Whittaker (1919).

[6] The unit of electric charge is named after Coulomb.

[7] See the article by Heering (1992).

[8] It is somewhat unfortunate that Franklin chose the "wrong" sign for the fundamental charge; it was later discovered that an electric current is due to the flow of *negatively* charged electrons (according to his convention), so that an electric current from A to B is physically a flow of electrons from B to A.

3.2 Electromagnetism

At the beginning of the nineteenth century, physicists had already realized that there
was a close relationship between electricity and magnetism. In Denmark, H. C.
Oersted (1777–1851) discovered that an electric current in a wire orients a magnet
perpendicularly to the current. When news of his observations reached France in
1820, Jean Baptiste Biot (1774–1862) and Felix Savart (1791–1841) repeated the
experiment and deduced the law that quantitatively describes the force exerted by an
electric current on a magnet. The magnet tends to be aligned in a direction perpendicular to the plane defined by the current and the magnet.[9] We are far from the
simplicity of the inverse square law!

Finally, the most in-depth study of how magnetism is related to the electric current
is due to André-Marie Ampère (1775–1836). Ampère discovered that two electric
currents repel or attract each other depending on whether they flow in the same or
opposite directions, something that reminded the properties of magnets. He reported
his results in a comprehensive treatise, *Mathematical Theory of Electrodynamical
Phenomena,*[10] published in 1826, in which he correctly deduced, from first principles
and his own experiments, the forces between electric currents. Ampère's great merit,
in addition to giving a mathematical description of these forces, was to realize that
magnetism is not a primary phenomenon, but an effect produced by the flow of
electricity. Accordingly, Ampère anticipated that magnetism is due to microscopic
electrical circuits in the molecules that make up matter, a conjecture that would be
fully confirmed with the atomic theory of the twentieth century.

To Michael Faraday (1791–1867), in England, we owe the most crucial experimental studies of electric and magnetic phenomena, reported in detail in a comprehensive treatise entitled *Experimental Researches in Electricity*, in three volumes
that appeared between 1839 and 1855. Among his most valuable contributions is the
discovery of electrical induction, which consists of the generation of electric currents
by moving magnets. This effect turned out to be the most efficient way of generating
electrical currents, much more efficient than voltaic batteries. It is the basis of all
modern technologies to produce electrical energy.[11]

It is noteworthy that Faraday always preferred a direct description of natural
phenomena in terms of images and models, and avoided mathematical reprentations
as far as possible, since he distrusted theoretical speculations. Aware of the insuffi-
ciency of a purely mechanistic representation of nature, he conceived the concept of
"lines of force" to explain how a magnet attracts iron: he imagined lines emanating
from the poles of a magnet and aligning iron filings, just as Peregrinus and Descartes

---

[9] In modern vectorial notation, the magnetic field produced by an element of current $i\,ds$ at position
r with respect to that element is the vector cross product: $i\,ds \times r/r^3$.

[10] André-Marie Ampère (1883), *Théorie mathématique des phénomènes électro-dynamiques:
uniquement déduite de l'expérience* (in French) (2nd ed.), A. Hermann.

[11] According to an anecdote, someone asked Faraday what the use of his discovery was, to which
he answered with another question: "What is the use of a newborn?".

had observed centuries before. Thus appeared the concept of something physical that cannot be seen or felt, but that has perfectly observable effects.

Another great contribution of Faraday was to chemistry, with his experiments on the dissociation of elements by electricity. He strongly suspected the existence of atoms, which would be related to electricity

> Although we know nothing of what an atom is, yet we cannot resist forming some idea of a small particle, which represents it to the mind; and though we are in equal, if not greater, ignorance of electricity, … yet there is an immensity of facts which justify us in believing that the atoms of matter are in some way endowed or associated with electrical powers, to which they owe their most striking qualities, and amongst them their mutual chemical affinity.[12]

Thus, he concluded that "the equivalent weights of bodies are simply those quantities of them which contain equal quantities of electricity…", so that "the atoms of bodies which are equivalent to each other in their ordinary chemical action have equal quantities of electricity naturally associated with them."[13] A century later it would be fully confirmed that the atoms of each chemical element are determined by the electrical charge of their nuclei.

3.3 Maxwell

A few years after Faraday's death, the great Scottish physicist James Clerk Maxwell (1831–1879) published his famous *Treatise on Electricity and Magnetism*, in which he reduced to mathematical formulas everything that was known in his time about electricity and magnetism: Coulomb's law, Biot-Savard's law, Faraday's induction, Ampère's theory… Maxwell's synthesis made it clear, beyond any doubt, that electricity and magnetism are two facets of one and the same interaction of nature, which can be called electromagnetism.

At the end of the second volume, in Chap. IX, Maxwell synthesized all the laws discovered by his predecessors into a set of partial differential equations, with which it was possible to calculate the electric and magnetic forces produced by any given distribution of charges and electric currents. Once again, partial differential equations entered physics to dominate it, as Einstein had noticed.

In the preface to his work, Maxwell distinguished between two ways of interpreting natural phenomena: one is Faraday's, which seemed to dispense with mathematics, and another, typical of continental scientists, which was based on them. However, Maxwell emphasized that the two methods were closely related, as he stated in a text that deserves to be quoted at length:

> I was aware that there was supposed to be a difference between Faraday's way of conceiving phenomena and that of the mathematicians, so that neither he nor they were satisfied with each other's language…

---

[12] *Experimental Researches*, § 852.
[13] *Experimental Researches*, § 869.

As I proceeded with the study of Faraday, I perceived that his method of conceiving the phenomena was also a mathematical one, though not exhibited in the conventional form of mathematical symbols. I also found that these methods were capable of being expressed in the ordinary mathematical forms, and thus compared with those of the professed mathematicians.

For instance, Faraday, in his mind's eye, saw lines of force traversing all space where the mathematicians saw centers of force attracting at a distance: Faraday saw a medium where they saw nothing but distance: Faraday sought the seat of the phenomena in real actions going on in the medium, they were satisfied that they had found it in a power of action at a distance impressed on the electric fluids.

When I had translated what I considered to be Faraday's ideas into a mathematical form, I found that in general the results of the two methods coincided, so that the same phenomena were accounted for, and the same laws of action deduced by both methods, but that Faraday's methods resembled those in which we begin with the whole and arrive at the parts by analysis, while the ordinary mathematical methods were founded on the principle of beginning with the parts and building up the whole by synthesis.[14]

We will return to this comparison of the two views of reality in the second book.

The electric and magnetic fields produced by a given distribution and flow of charges can be calculated directly from Maxwell's equations. However, for the reverse problem, to calculate the motion of a charged particle in a given electromagnetic field, it is necessary to complement Maxwell's theory with an additional equation for the force exerted by the field on the particle, which is called the Lorentz force.[15] It can be said that the Lorentz force was the last force in the Newtonian sense—change of momentum—to appear as such in physics. It is remarkable that the Maxwell field and the Lorentz force fit very naturally into the Lagrangian and Hamiltonian formalisms, making it clear that these formulations of mechanics have a much wider range of application than their creators imagined. Also, as we will see in the next Chapter, Maxwell's and Lorentz's equations appear naturally in the theory of relativity and, in fact, they were the key to the formulation of this theory.

3.4   The Field

Michael Faraday introduced the notion of field when he noticed that electrically or magnetically charged bodies interact through invisible and immaterial "lines of force". Inspired by Faraday, Maxwell proposed the concept of field, which he defined as a form of "energy without matter". Like Newton's gravity, the field is invisible, but its effects on material objects are perfectly real and can be well described by mathematical formulas.

In the mathematical theory developed by Maxwell, electric and magnetic fields are represented by two vectors that define their directions and magnitudes.[16] In general,

---

[14] *A Treatise…* introduction to the first edition.

[15] Lorentz will be introduced in the next chapter, in the context of the theory of relativity.

[16] A vector can be thought of as an arrow in space. The electromagnetic field can be represented as a set of arrows that fill the entire space and whose sizes and directions depend on each point.

physicists nowadays define a field as a mathematical function of space and time, that is, a function that assigns a certain quantity or set of quantities (for instance, the three components of a vector) to each point in space-time.

With his theoretical synthesis, Maxwell achieved for electromagnetism what Newton did for gravity, although Maxwell's equations turned out to be much more complicated than those of Newton (explicitly eight equations for the six components of the electric and magnetic vectors!). However, unlike his illustrious predecessor, Maxwell did not concern himself with the problem of action at a distance, as the thinkers of previous centuries would have liked. Instead, his concept of the field as something different from matter, but with physical effects, turned out to be fundamental in twentieth century physics due to a very wide range of applications, including gravity itself (one could conceive a "gravitational field!", as we will see in the next chapter). Although, in a certain sense, the Maxwellian field is reminiscent of the Ether, it is perfectly well defined mathematically.

The field was fully accepted by the physicists of the twentieth century. In this regard, Einstein noted:

> ... [T]he concept of the material object was gradually replaced as the fundamental concept of physics by that of the field. Under the influence of the ideas of Faraday and Maxwell the notion developed that the whole of the physical reality could perhaps be represented as a field whose components depend on the four space-time parameters.[17]

3.5 Light

One of the great successes of Maxwell's theory was the explanation of the nature of light. In that regard, let us briefly recall that there were originally two theories. According to Huygens in his *Treatise on Light*, it would be a wave which vibrates in some very subtle and impalpable medium—again the Ether. According to Newton, as extensively described in his *Optiks*, light would consist of particles traveling in straight lines, bouncing off objects or penetrating transparent media. It was in the very beginning of the nineteenth century that the question seemed to be settled by Thomas Young with his classical experiment demonstrating that light exhibits the phenomenon of *interference* (when two waves meet, their amplitudes add in certain positions and cancel at others, producing a pattern of bright and dark fringes[18]). The experimental confirmation seemed irrefutably in favor of Huygens and against Newton.

In case there would still be some doubts about the nature of light, Maxwell proved, from his equations, that it is an electromagnetic wave: a joint vibration of electric and magnetic fields, vibrating *perpendicularly* to the direction of the propagation (a direction that defines the *polarization* of light), in contrast with sound waves whose vibrations are *longitudinal,* i.e., in the direction of propagation. In the general case

---

[17] Jammer (1954), foreword.

[18] More on the subject in the chapter on quantum mechanics.

of a fluid, the equation that describes the propagation of a wave contains the speed of sound in the medium as a fundamental parameter.[19] In the equivalent equations obtained by Maxwell, the speed of light appears directly as the speed of the wave. There was no doubt that light is an electromagnetic wave. Thus, modern optics was born as a special branch of electromagnetism.

Since, in principle, there can be waves with any wavelength,[20] it was to be expected, according to Maxwell's theory, that "light" would exist in Nature with wavelengths both larger and shorter than that of visible light. And indeed, there is an entire spectrum of electromagnetic radiation, from radio waves to gamma rays, including infrared, visible light, ultraviolet, and X-rays.

3.6   Ether Again?

After Faraday and Maxwell, physicists definitively gave up trying to explain phenomena like electromagnetism and gravity by recurring to some material mechanism. Lord Kelvin was perhaps one of the last great physicists to follow the mechanistic tradition until the dawn of modern physics. Still at the beginning of the twentieth century, he stated that to understand a physical process, he was not satisfied until he conceived a mechanical model that explained it: "If I succeed, I understand; otherwise, no."[21]

Nonetheless, the question remained on how the electromagnetic interaction is transmitted and in what medium light propagates. Apparently, there was no other choice than to postulate, as the ancient philosophers had, the existence of an Ether: an invisible and impalpable substance that would be the support of electromagnetic phenomena. For some time, the most distinguished physicists discussed and speculated on the subject, proposing various hypotheses, each more ingenious, about the nature of this mysterious entity: Was it like a fluid or like a gas? Was it compressible or incompressible? Could it drag bodies? Was it fixed with respect to the Earth or to the Universe? But they could never reach an acceptable conclusion. Finally, as we will see in the next chapter, such a problem was simply abandoned. As Paul Feyerabend[22] once remarked, the fundamental problems of science are never solved, but rather dissolved. It is necessary to wait for a new generation of scientists for the old problems that worried their elders lose their interest and become irrelevant.

---

[19] The analogous problem of a vibrating string was studied by D'Alembert, Euler, Bernoulli and Lagrange. See the article by Wheeler and Crummett (1987).

[20] A wave is characterized by its wavelength λ, which is the distance between one crest and the next, or equally by its frequency v, which is the number of times it vibrates per second. In the case of light, the product $\lambda v = c$ is the speed of light. The shorter the wavelength, the higher the frequency, and vice versa.

[21] Lord Kelvin, *Baltimore Lectures on Molecular Dynamics and the Wave Theory of Light* (1904).

[22] Feyerabend (1975), *op. cit.*, particularly Chap. XV.

Chapter 4
Modern Physics: Relativity

Abstract Maxwell's electromagnetic theory was the basis of Einstein's special theory of relativity. This theory appeared to be most natural in a four-dimensional space-time, as proposed by Minkowski. For a theory of gravity, Einstein had the brilliant idea of generalizing this four-dimensional space to a curved space, as had been previously studied by Riemann. Hilbert followed a similar path, based on the principle of minimal action. Since then, the theory has been confirmed in many ways.

The two theories that revolutionized physics in the twentieth century, relativity and quantum mechanics, have very different structures, motivations, and methods. Einstein's theory of relativity, both in its special and general versions, is impressive for the clarity of its concepts and the impeccable logic of his reasoning. Quantum mechanics, on the other hand, is based on incomprehensible and unclear notions, and its mathematical basis (despite many worthwhile attempts) is only justified by its impressive results. Surely the greatest mystery of quantum mechanics is that it works… and it does it extraordinarily well! Let us summarize below the basic ideas underlying these two pillars of modern physics.

4.1 Relativity: From Galileo to Einstein

If the Earth moves around the Sun, why is this motion not perceived? To defend the heliocentric system, Galileo argued that in a reference system in uniform motion (in a straight line and with constant velocity), it is impossible to detect the motion locally by means of physical experiments.[1] This is known as Galileo's principle of relativity: the laws of nature have the same form in any inertial reference system - that is, in which there are no inertial forces such as those produced by acceleration or deceleration, or centrifugal forces due to a curved path—. In a vehicle in straight

---

[1] Strictly speaking, the Earth rotates and does not move in a straight line, but the centrifugal acceleration due to these effects is practically imperceptible.

© The Author(s), under exclusive license to Springer Nature Switzerland AG 2023   39
S. Hacyan, *The Mathematical Representation of Physical Reality*,
The Frontiers Collection, https://doi.org/10.1007/978-3-031-21254-3_4

and uniform motion—viz. a train, an airplane, etc.—, it is impossible to determine if it is moving without looking outside.

In mathematical terms, a Galileo transformation is a change of the spatial coordinates (for instance, the Cartesian coordinates x, y, z) with which one goes from one inertial reference system to another moving with constant speed with respect to the first. In classical physics, time is an absolute concept since it is the same in all systems: all the clocks show the same time… or so we suppose. Likewise, velocities are simply added or subtracted depending on which reference frame they are measured in.

But what about a ray of light? Being an electromagnetic wave, it must have the same speed with respect to the Ether (if it exists!) and, consequently, it would be possible to detect the motion of the Earth through that universal medium. Galileo's principle of relativity would not apply to electromagnetic phenomena.

A purely mathematical way of checking this is to note that Maxwell's equations do change their form by simply moving from one reference frame to another by means of a Galileo transformation of the coordinates. It was thought, therefore, that these equations should be valid in only one particular system, which would be obviously the system in which the Ether is at rest, and that these same equations would have a somewhat different form in another reference system, such as the Earth in orbit around the Sun. Thus, the velocity of light would depend on the velocity of the Earth with respect to the Ether, and such a slight difference should be detectable experimentally. The famous experiment of Michelson and Morley,[2] in 1887, was intended to measure the speed of light emitted in different directions with respect to the Earth's orbit. Contrary to what was expected… no speed difference was detected!

As physicists clarified the situation, Hendrik A. Lorentz (1853–1928) decided to investigate Maxwell's equations with a purely mathematical approach. If the form of these equations changes under a Galilean transformation, what kind of coordinates transformation would not change their form? Lorentz found the solution in a kind of transformations that do not only modify spatial coordinates, but also time, as if time were just another coordinate. Around the same time, Henri Poincaré reached the same result, making it even more general by including translations in space and time. But Lorentz and Poincaré's treatments were considered just mathematical curiosities.

In 1905, young Albert Einstein (1879–1955) published a total of five articles that would revolutionize the physics of the new century. One of these articles dealt precisely with the incompatibility of Maxwell's equations with Galileo's principle of relativity. As if cutting a theoretical Gordian knot, Einstein ignored the Ether and postulated that the speed of light is the same in all reference frames. Starting from this bold, but very simple postulate, Einstein showed that both space and time must depend on the motion of the observer. After all, speed is distance divided by time, and if distance *and* time contract in the same proportion for light, the speed of light remains constant. Einstein developed the mathematical consequences of

---

[2] Albert A. Michelson (1852–1931) and Edward W. Morley (1838–1923), American physicists. The former received the Nobel Prize in Physics in 1907 for his research with interferometers, which are very high-precision optical instruments.

his hypothesis. A major part of the article is devoted to showing that Maxwell's equations do not change in form under the transformation of the spatial and temporal coordinates previously discovered by Lorentz (who, by the way, Einstein did not mention in his 1905 paper, he later argued that he was unaware of his elder's work). Thus, the principle of relativity proposed by Einstein, which generalizes that of Galileo's, made it unnecessary to invoke the mysterious Ether and, by the same token, explained the negative result of the Michelson-Morley experiment (an experiment that Einstein did not mention either).

Thus, the speed of light turned out to be a fundamental constant of nature, which in modern physics is designated by the letter c.[3] The basic principle, to be the same for all observers, seems to contradict the common concept of speed. However, speeds in the theory of relativity do not simply add or subtract.[4] In fact, relativistic effects such as time contraction are noticeable only at speeds very close to the speed of light. For lower speeds, such as those we are used to in our daily experience, the effect is so small that it can only be measured with extremely precise instruments.

One of the most important predictions of the theory of relativity is the relationship between mass and energy, expressed by the famous formula $E = mc^2$, where E is the energy, m the mass and c the speed of the light. If energy, in the nineteenth century, seemed to be an abstract concept and only a useful mathematical tool, it acquired a new relevance in modern physics and became one of the most fundamental concepts, at the same level as mass. Einstein's formula, in fact, is valid for a body at rest if m is interpreted as its "rest mass", that is, the mass measured in a reference frame in which the object to be weighed is at rest. For a moving body, the correct formula for the energy is $E = m c^2 \gamma$, where $\gamma = 1/\sqrt{1 - v^2/c^2}$ is the so-called Lorentz factor.[5] As a consequence, the speed of light is an insurmountable natural limit, since a material body would need an infinite amount (!) of energy to reach that speed. Only the photon, which has no mass but only energy (according to all experimental evidence) can move at the speed of light. To date, no physical evidence has been found that this theoretical constraint does not hold (although there are many attempts in the domain of science fiction).

It is interesting to notice that Einstein's original paper did not mention time as a fourth coordinate. The precise mathematical concept of a space-time of four dimensions—three spatial and one temporal coordinates—, is due to Herman Minkowski (1864–1909), who presented his idea in a 1908 paper. Minkowski showed that relativistic physics can be formulated in terms of a four-dimensional geometry, in which time is interpreted as a fourth coordinate[6] and each "point" must be interpreted as an event occurring at a given place at a given time. Each point of that four-dimensional

---

[3] From Latin *celeritas*.

[4] The relativistic formula for the "addition" of velocities is $(v_1 + v_2)/(1 + v_1 v_2 /c^2)$. The denominator is practically 1 for very low speeds compared to c (the speed of light), which is why it goes unnoticed in our everyday experience.

[5] In some textbooks, $m \gamma$ is defined as the "moving mass" and m as the "rest mass".

[6] Originally, time was interpreted as an imaginary coordinate, but nowadays it is preferred to consider it as real and include a negative sign in the formula to measure "distances" (see next note).

space is represented by a set of four numbers, three of which locate the position of an event in space, and the fourth coordinate measures the time at which it occurs. Additionally, a formula is needed to measure the "distances" between point-events, and this is given by a generalization of the Pythagoras theorem.[7] Minkowski's formulation turned out to be the most natural (elegant, as physicists and mathematicians say) way of describing relativistic physics, and in particular the electromagnetic theory: the cumbersome equations of electrodynamics as originally proposed by Maxwell take a very simple and compact form in Minkowski's four-dimensional space-time notation. Relativity thus turned out to be a tailor-made theory for electromagnetism. It also revealed an unexpected mathematical connection between space and time.

4.2   General Relativity (1)

After the success of his 1905 theory, supplemented with Minkowski's geometry, Einstein spent a long time pondering on how to extend the principle of relativity to systems that are not inertial (that is, systems in which there are inertial forces produced by changes in velocity or trajectory). After many years of trials and errors, Einstein presented in 1915 the final version of the theory of relativity that includes the force of gravity. A basic postulate of the theory was the "principle of equivalence", according to which, as Galileo had noticed, all bodies fall with the same acceleration regardless of their mass. This is due to the exact proportion between the inertial mass, which is the resistance of a body to change its state of motion, and the gravitational mass, which is determined by the force of gravity.

Einstein, in his own words, found the key to the problem of gravity when it occurred to him that if one is falling from the roof of a house, the force of gravity is not felt. This idea is often illustrated by the example of an observer trapped in an elevator in free fall: the observer would float without apparent gravity, just like all the objects around it, and might believe to be in outer space, far from any attracting planet.[8] However, if such an elevator were large enough, the gravitational attraction would not act on it uniformly due to tidal forces (that is, because the gravitational attraction decreases with distance and therefore different parts of an extended body are not attracted evenly). In other words, it is possible to "cancel" gravity in a small local environment, but not in a large enough region.

All the above suggested Einstein an ingenious analogy with geometry. We know that Euclidian geometry applies on a flat surface. For instance, two parallel lines, by their very definition, always remain parallel no matter how long they are extended. However, on a curved surface, Euclidean geometry breaks down because there are

---

[7] In three dimensions, the distance from the origin O to the point with coordinates (x, y, z) is $x^2 + y^2 + z^2$. In Minkowski space, the "distance" from the origin to the event-point (x, y, z, t) is $x^2 + y^2 + z^2 - c^2 t^2$ (notice that it can be positive, negative, or zero due to the minus sign in front of the time coordinate).

[8] For instance, astronauts in orbit around the Earth are in a free fall... that never reaches ground.

no straight lines on a curved surface. This is evident on the surface of the Earth: a meridian can be thought of as the equivalent of a straight line, but two "straight lines" so defined, initially "parallel", would end up meeting at the Poles. Nevertheless, the surface of the Earth appears to be flat at a small scale.

Similarly, it occurred to Einstein that space-time should look like Minkowski's space-time to an observer in free fall because he does not perceive the existence of a gravitational field in his immediate environment. However, at a large scale, gravity does manifest itself and it must be related to... a curvature of space-time! Minkowski's space-time would be the four-dimensional equivalent of flat space, while gravity would be the manifestation that space-time is curved.

But what is a four-dimensional curved space-time? The only curved space that we can visualize is a two-dimensional surface: for instance, the surface of the Earth or any curved body. Fortunately, nineteenth-century mathematicians had already dealt with the problem of spaces with any number of dimensions, not only two or three. For the general version of his theory of relativity, Einstein took advantage of the new geometry created a few decades earlier by the great mathematician Bernard Riemann (1826–1866).

But first, a clarification: why can the concept of a curved space be used to describe gravity, but not electromagnetism, which is the other fundamental (macroscopic) interaction of Nature? Why is there no geometric analogy for the electromagnetic field? The reason is the equivalence principle mentioned before, according to which bodies fall with the same acceleration independently of their mass. Due to this fact, a single curved space-time can be associated to a given gravitational field, and the motions of all the bodies in that field (or space) will be independent of their inertial mass. In contrast, there is no such equivalence for electricity: the inertial mass is totally independent of the electric charge. Therefore, for electromagnetic forces, it would be necessary to define a different curved space-time for each charged particle, according to its very particular ratio of mass to electric charge... which would make no sense because it would break the universality of the formalism.[9]

4.3 Beyond Euclid

Before proceeding with gravity, let us open a parenthesis on geometry. Euclid's *Elements,* which was to become the model of all studies on logic and mathematics, begins with the definitions of primary geometric objects such as points, lines, circles, etc., and some postulates they must satisfy. The first four postulates of the *Elements* are very clear: (1) A straight line passes through two points; (2) A finite straight line can be extended by another straight line; (3) A circle can be constructed with any

---

[9] One (unsuccessful) attempt to geometrize electromagnetism was to generalize space-time to a fifth (!) dimension that would include this interaction: the Kaluza-Klein theory.

center and radius; (4) All right angles are equal. Finally,[10] appears the fifth postulate in a very confused form which contrasts with the previous ones, and that is basically equivalent to stating that only one line parallel to another line can pass through a given point.

The fifth postulate might seem obvious, so much so that it should be possible to deduce it from the previous four just as another theorem of geometry. Over several centuries, many mathematicians tried to prove it from the previous postulates, but without any success. Finally, instead of stubbornly searching for a proof, the mathematicians Janos Bolyai (1802–1860), on the one hand, and Nikolai I. Lobachevsky (1792–1856), on the other, decided to attack the problem from a different direction: what happens if that postulate is dispensed with? If some contradiction arises, that would be a proof of the fifth postulate by *reductio ad absurdum*. However, both Bolyai and Lobachevsky reached the conclusion that new geometries, non-Euclidean but perfectly consistent, could be created with their own postulates and theorems, depending on what is defined as a "straight line". For instance, on the surface of a sphere, the closest thing to a straight line is a maximum arc, but two arcs cannot be parallel: the Bolyai geometry is that of a spherical surface, like that of Earth. Lobachevsky's geometry corresponds to the surface of a saddle: by definition, at least two "straight lines" can pass through a given point that are "parallel" to a given "straight line".[11]

In that same context, Riemann developed an important generalization of Euclidean geometry that includes not only two-dimensional curved surfaces—easy to visualize—, but also "curved spaces" with any number n of dimensions, which are impossible to conceive except through mathematics. A two-dimensional surface (such as the Earth) requires two numbers (longitude and latitude) to locate a point. All that is needed, remarked Riemann, is to identify a "point" with a set of n numbers, and to define a formula for measuring distances between two "points" that would generalize the Pythagoras formula. In three dimensions, for instance, the square of the distance between two points with Cartesian coordinates (x, y, z) and $(x + dx, y + dy, z + dz)$ is $dx^2 + dy^2 + dz^2$. In a more general Riemannian space, the equivalent formula for measuring "distances" between two points is given by the formula

$$ds^2 = g_{xx}dx^2 + g_{yy}dy^2 + g_{zz}dz^2 + 2g_{xy}dxdy$$
$$+ 2g_{xz}dxdz + 2g_{yz}dydz,$$

---

[10] In its original version, the fifth postulate was: *If a line crosses two lines in such a way that the interior angles on the same side add up to less than two right angles, the two lines extended indefinitely intersect on the side where the two angles add up to less than two straight* It was Proclus, a fifth-century Byzantine philosopher and mathematician, who showed that it was equivalent to the simpler form mentioned here.

[11] In technical mathematical language, the Bolyai space has a positive curvature, and the Lobachevsky space has a negative curvature.

where g_{xx}, g_{yy}, etc. are known in technical language as the components of the *metric tensor* g_{ij} (the subscripts i and j run from 1 to n, the number of dimensions).[12] Accordingly, a Riemannian space of n dimensions is uniquely defined by means of its metric tensor. As for the concept of straight line, it must be replaced in a curved space by that of *geodesic*, a curve defined as being of minimum length between two points (for instance, a meridian on the Earth's surface).

Riemann further developed the mathematical apparatus of his geometry in another work on... the thermal properties of solids! There appeared the "Riemann tensor", which is constructed with first and second derivatives of the metric tensor, and which indicates whether a space is really curved or not.

4.4 General Relativity (2)

Based on Riemannian geometry, Einstein developed a generalization of relativity to gravitational forces. His basic postulate was that space-time is a four-dimensional Riemannian space, and the gravitational force is a geometric property of space–time, thus successfully unifying physics with geometry. Space, combined with time, was no longer a simple scenario of natural phenomena, as in Newtonian physics, but with a dynamic role. Likewise, the geodesics in that space are interpreted as the trajectories of particles in a gravitational field.

Einstein, with the help of his friend the mathematician Marcel Grossman (1878–1936), deduced the fundamental equations that relate a distribution of mass and energy to the gravitational field it produces, described as a Riemannian space. These are partial differential equations (of course!) for the ten components of the metric tensor g_{ij} that define a curved space-time in four dimensions.

The equations of general relativity are to gravitation what Maxwell's equations are to electromagnetism, although they are considerably more complicated.[13] In Maxwell's theory, the electromagnetic field is described by six quantities that correspond to the components of the electric and magnetic fields. These fields, in turn, can be described by four functions, which are their potentials, and which satisfy four partial differential equations of second degree, related to electric charges and currents.[14]

As for the gravitational field, it is described by the ten components of the metric tensor that act as potential functions. From its first and second partial derivatives, the so-called Riemann tensor is constructed, which, in a four-dimensional space, has twenty components that uniquely describe a curved space–time. These components

---

[12] The metric tensor is symmetric since $g_{ij} = g_{ji}$. In three dimensions, it has 6 independent components, and in four dimensions, 10 independent components.

[13] Technically, the electromagnetic field is a spin 1 field, and the gravitational field is a spin 2 field (see Chap. 6).

[14] In modern notation and in Minkowski space-time, the electromagnetic field is described by an antisymmetric tensor F_{ij} given by first partial derivatives of a potential vector with four components A_i.

are, in a sense, the equivalent of the six components needed to describe an electromagnetic field. For a certain combination of the components of the Riemann tensor,[15] Einstein deduced a set of equations that determine the gravitational field produced by a given distribution of mass and energy.

Explicitly, the Einstein equations are ten partial differential equations that relate geometry with matter. One side of these equations is given in terms of the metric tensor, that is, geometry. The other side represents matter and energy. They are nonlinear equations, which means that gravity interacts with itself, unlike electromagnetism.[16] Due to this complication, only some exact solutions are known for very particular physical cases. It is only in recent years, with the development of very powerful computers and programming methods that numerical solutions have been obtained for a variety of cases, mainly in astrophysics.

Einstein, in his original article, had to use approximate solutions of his equations, valid for weak gravitational fields. He was able to predict and explain two astronomical effects: the shift of the perihelion of the orbit of the planet Mercury and the deviation of a ray of light passing close to the Sun. The former effect was already known by the astronomers of the nineteenth century: Mercury's orbit is not a closed ellipse, but its perihelion advances about 9.55 min of arc per century. This effect had been impossible to explain with Newtonian theory, but Einstein's theory described it with great precision.

The other effect is the bending of a light ray as it passes near a very massive object like the Sun. The first successful observation of this effect is due to the British astrophysicist Arthur S. Eddington (1882–1944). Eddington took advantage of a total solar eclipse in Africa to take pictures of the stars near the Sun and compare their positions with those they had six months earlier without the Sun nearby. In November 1919, in a solemn session of the Royal Society in London, Eddington announced that, as seen in his photographic plates, the displacement of the stars had been observed as predicted by Einstein. On that day, Einstein rose to fame. Since then, all the predictions of general relativity have been verified with great precision; so far, no discrepancy with observations has been found.

As for particular but rigorously exact solutions of the equations of general relativity, the first such solution was found by the German astronomer Karl Schwarzschild (1873–1916), only a few months after the publication of Einstein's seminal article. The Schwarzschild solution describes the gravitational field produced by a spherical symmetric mass distribution.[17] Its most remarkable property is that the corresponding space-time curves until it closes on itself if the mass is concentrated in a certain radius, the Schwarzschild radius $2GM/c^2$, where M is the mass of the sphere.

---

[15] The Ricci tensor R_{ij} with ten components in a four-dimensional space.

[16] A linear equation can have several solutions and the sum of any of its solutions is also a valid solution. Maxwell's equations are linear: the electric field of two distinct charges is equal to the sum of the two separate fields, and two light rays do not interact with each other.

[17] Schwarzschild died shortly after writing his article due to an infection contracted on the Russian front, where he was fighting for his country during the First World War.

For a star with the mass of the Sun, this radius is equivalent to about 3 km and determines the so-called *event horizon*, a closed surface that light can only penetrate but not leave. Physically, it would be a body so massive that the escape velocity from its surface would exceed the speed of light. Curiously, Laplace had already noticed this possibility with only Newtonian physics, and had even speculated that "dark bodies" could exist in the Universe with a gravitational field so intense that light could not escape from them. Currently, astronomers have found strong evidence that objects with these characteristics, which were theoretically predicted, exist throughout the Universe: they are now called "black holes".

Shortly after Schwarzschild's paper, another exact solution of Einstein's equations appeared that describes a body with mass and also electric charge. The Reissner-Nordström space–time[18] has two event horizons, one inside the other. In theory, a particle (or a spaceship!) could penetrate the outer horizon, then the inner horizon, then exit both horizons and emerge... in a parallel universe! Of course, parallel universes exist only in the mathematical world, and it is not obvious that they correspond to anything in the real world. Similarly, in the 1960s, Roy Kerr (a New Zealander physicist) found another exact solution of Einstein's equations that represents a rotating black hole, also with two event horizons. It then took a tremendous effort for the mathematical physicists to prove that the most general solution of Einstein's equation that represents a black hole (that is, an object with an event horizon) is the electrically charged Kerr solution.[19] In this mathematical sense, a black hole would be like an elementary particle, since it is characterized by only a few parameters: mass, "spin" and possibly electric charge.

Another very important application of the general theory of relativity is to cosmology, the study of the Universe at a large scale, which would have been impossible without this theory. We will see this topic in Chap. 8.

4.5 Hilbert

It is little known that the equations of general relativity that Einstein deduced were also obtained by the great mathematician David Hilbert (1862–1943) using... the principle of least action! Both scientists formally presented their results to their colleagues, Einstein at the Berlin Academy and Hilbert at his Göttingen Institute, in November 1915, Hilbert five days before Einstein.[20]

However, the two followed very different methods. Einstein relied on his physical intuition to "guess" what form the fundamental equations should have. Additionally, he established a method of approximation to solve his equations and thus obtained

---

[18] H. J. Reissner (1874–1967), German engineer. G. Nordström (1881–1923), Finnish physicist.

[19] Known as the Kerr-Newman solution, discovered by American physicist E. T. Newman (1929–2021) and his students.

[20] Mehra (1974).

two physically meaningful results: the perihelion shift of Mercury and the bending
of light by a gravitational field.

As for Hilbert, he fully accepted Einstein's idea of interpreting the gravitational
field in terms of a Riemannian geometry, but he followed his own path. As he stated,
he was inspired by Gustav Mie's attempt[21] to develop a theory of electromagnetism
more general than Maxwell's (more on Mie's theory in Chap. 15).[22] The idea was to
identify a scalar function (that is, a function that has the same value in any coordinates
system) and use it as a "world function", actually what we now call a Lagrangian
function. With that function at hand, the next step was to define an action and apply
the principle of least action to obtain the equations that describe the physical system.
Hilbert identified the gravitational Lagrangian with the Ricci scalar, that function of
the components of the metric tensor g_{ij} and their first and second derivatives. Thus,
the representation of the physical world was reunited with geometry and with the
principle of least action!

There was never a priority dispute between the two scientists who admired each
other. Hilbert built on preliminary ideas that Einstein had published a few years
before his seminal paper on the gravitational field. And although Einstein never gave
him any credit, Hilbert had enough of merits of his own to go down in history as the
great mathematician of the twentieth century.

The crucial point of Hilbert's formalism was that the action should be a scalar,
that is, it must remain invariant under a change of reference system. Otherwise, the
action could be minimal in one system but not in another, which would make no
sense. In turn, the invariance of the action, defined as the integral of the Lagrangian
(by the Hamilton principle, see Chap. 2) implies certain mathematical conditions
on the Lagrangian itself. This way of handling a problem of mathematical physics
would turn out to be very useful in relativistic and quantum mechanics, as we shall
see in the following chapters.

4.6   Energy and Gravitational Waves

Eddington, who confirmed the bending of light, was a great promoter of the relativistic
theory of gravitation and was particularly interested in its mathematical aspects. He
published in 1923 an extensive and detailed treatise on the mathematical aspects of
the theory of relativity and its physical and astronomical implications. One of the
problems he noted is the lack of a precise and consistent definition of energy in the
new theory.

> As soon as the principle of conservation of energy was grasped, the physicist practically
> made it his definition of energy, so that energy was that something which obeyed the law

---

[21] German physicist (1868–1957).
[22] Smeenk and Martin (2007).

of conservation. He followed the practice of the pure mathematician, defining energy by the properties he wished it to have, instead of describing how he had measured it.[23]

Indeed, to define unambiguously the energy of the gravitational field is still an open problem. The formula for the energy of the electromagnetic field is well known since the time of Maxwell,[24] but there is no equivalent formula for the gravitational field. The most Eddington could get was a very complicated formula, the so-called *energy pseudo-tensor*, that involved the spatial and temporal derivatives of the metric tensor g_{ij}. However, Eddington's formulation, as he himself was clear about, had the defect of depending on the coordinate system, so that the energy could be made zero at each point just by changing to the appropriate coordinate system (that is why he named it pseudo-tensor and not tensor). It was a violation of the so cherished condition in theoretical physics that the laws of nature must not depend on the particular reference system… but there was apparently no choice.

Eddington only outlined the procedure for calculating the pseudo-energy tensor. A decade later, Richard Tolman[25] published his own treatise[26] on the theory of relativity and, among other things, deduced the explicit form of the pseudo-energy tensor in terms of the metric tensor g_{ij}. It is an impressive formula that consists of no less than… sixteen terms! Nothing to do with the simplicity or "elegance" of the important formulas in physics.

The problem of defining energy is clearly manifested in the phenomenon of gravitational radiation.[27] The year after he published the final version of his theory, Einstein showed that, in the approximation of weak gravitational fields he had used (i.e., small curvature of space-time), his equations reduced to a wave equation, similar to the equation for electromagnetic waves in Maxwell's theory, with the speed of light as the speed of the gravitational waves. This result provoked many discussions that were to last several decades: do gravitational waves really exist or are they mere mathematical constructs? Einstein himself had doubts about their existence and, even if they were real, the possibility of detecting them. Eddington referred to gravitational waves as "travelling at the speed of thought".[28] The snag is that gravitational waves should carry energy just like electromagnetic waves, and thus produce some motion of matter in their wake, but there is no clear way to define energy in curved spaces.

In the 1950s, the English translation of the monumental "Landau and Lifshitz" *Course of Theoretical Physics* appeared.[29] In the chapter devoted to the theory of general relativity,[30] the authors used Tolman's lengthy formula for the energy pseudo-tensor (without giving him the credit) and deduced, as a particular exercise, what

---

[23] Eddington (1923). *The mathematical theory of relativity.* Chap. IV.

[24] The electromagnetic energy density (in Gaussian units) is $(E^2 + B^2)/4\pi$, where E and B are the electric and magnetic field vectors respectively.

[25] American physicist (1881–1948).

[26] R. Tolman, *Relativity, Thermodynamics and Cosmology* (1934).

[27] For a history of gravitational waves, see Kennefick (2007).

[28] Kennefick, *op. cit.*

[29] By the Russian physicists Lev D. Landau (1908–1968) and Evgeny M. Lifshitz (1915–1985).

[30] Landau and Lifshitz, *The Classical Theory of Fields.*

would be the energy radiated in the form of gravitational waves by a binary system of two stars rotating round each other. Due to the consequent loss of energy of the system, the distance between the two stars should decrease and they would end up collapsing against each other.

The validity of the formula obtained by Landau and Lifshitz was very much debated at the beginning, due to the lack of a mathematically rigorous definition of energy. However, it was spectacularly confirmed in 1974 with the discovery of a binary pulsar,[31] a system composed of two neutron stars revolving around each other in very close orbits. Thanks to the impressive precision of the observations, it was possible to verify that the Landau and Lifshitz formula for the energy loss of the binary system was satisfied with extreme precision and beyond any doubt.[32] Since then, many more binary systems have been discovered that fully satisfy the theoretical prediction of Landau and Lifshitz. Physicists no longer had any doubts that gravitational waves exist, and so much so that large detectors of these waves were built to open a new window for the observation of the Universe. The 2017 Nobel Prize[33] was precisely for the direct detection of gravitational waves in a ground-based observatory… but that's another story.

4.7   Beyond General Relativity

There have been various attempts to generalize Einstein's theory of gravity to more complicated formulations, for instance, supposing a hypothetical variation of the gravitational constant or other fundamental constants of nature. Attempts have also been made to extend Hilbert's method to actions more general than the Ricci scalar, with the result that equations more complicated than Einstein's original ones have been obtained. The main purpose of these attempts is to seek an alternative explanation for astronomical observations (to be mentioned in the following chapters). However, none of these attempts has reached a convincing result.

It should also be noted that the formulation of a theory of gravitation in terms of Riemannian geometry, while very convenient, is not the only possible one. The fundamental assumption of general relativity is that distances are measured with the basic formula that generalizes the Pythagorean theorem, a formula that includes (as we saw before) only double products of the differentials of length, like dx^2, or $dx\, dy$, etc. However, even more general geometries can be constructed in which the "distance" between two points is measured by a totally general function of the separation between points. In this sense, a possible geometry more general than

---

[31] A pulsar is a neutron star, the remnant of the core of a star that exploded as a supernova. Its size is only a few tens of kilometers, and its density is that of an atomic nucleus. Having an extremely high magnetic field, a pulsar emits electromagnetic waves that are detected as periodic radio pulses which can be measured on Earth with great precision.

[32] For the discovery of the binary pulsar, R. A. Hulse and J. H. Taylor received the Nobel Prize in 1993.

[33] R. Weiss (1932–), K. S. Thorne (1940–), and B. Barish (1936–).

Riemann's, is the Finsler geometry[34] in which there is no Pythagorean theorem, but something more general: the length of the hypotenuse of a right triangle would be a general function of its legs. The problem, however, is to obtain something equivalent but more general than the standard Einstein equations in such a geometry.[35]

---

[34] Paul Finsler (1894–1970), Swiss-German mathematician.

[35] See, for instance, F. Rutz, (1993): *A Finsler Generalization of Einstein's Vacuum Field Equations.*

Chapter 5
Modern Physics: Quantum Mechanics

I must confess I am jealous of the term atom, for though it is very easy to talk of atoms, it is very difficult to form a clear idea of their nature ...
Michael Faraday (Experimental Researches in Electricity, § 869.)

Abstract Planck proposed the quantization of energy as a mathematical trick to explain the black body spectrum. It was Einstein who postulated this quantization as a physical property of Nature. Following this idea and the newly discovered structure of the atom, Bohr proposed a model for the hydrogen atom. The mathematical basis of quantum theory began to be established by Heisenberg, Born and Jordan, who developed "matrix mechanics". Other quantized quantities, angular momentum and spin, appeared in their formalism. Then, Schrödinger deduced his famous equation which is the basis of (non-relativistic) quantum mechanics. This equation defines a "wave-function" whose interpretation has been the subject of endless discussions, viz., Bohr's Copenhagen interpretation.

By the end of the nineteenth century, most physicists, if not all, were convinced of the existence of atoms as the smallest constituents of matter. Their structure, however, was a matter of discussion. In this context, the discovery of the electron as an elementary particle, carrier of a fundamental electric charge—which turned out to be negative, according to the accepted convention—was crucial. Such a discovery is due to Joseph John Thomson (1856–1940) who experimented with cathodic rays, which are flows of free electrons, as we presently know. It became clear that the electric current is due to these particles, which are totally identical to each other, with the same charge and mass.[1] Thus, it was established that the electric charge is quantized: it can be counted one by one with the integer numbers.

---

[1] Unlike Leibniz's monads, which had to be totally different from each other (*Monadologie*).

5.1 In Illo Tempore

The birth of quantum mechanics is usually associated with the publication, in 1900, of a famous work by Max Planck on the energy emitted by hot bodies. Planck deduced a formula that correctly describes this energy as a function of the frequency[2] radiated inside a "black body"—basically, a closed oven in thermal equilibrium at a fixed temperature. After many attempts, Planck realized that he had no choice but to resort to the hypothesis that the molecules that make up the inner walls of the oven have quantized energies, that is, integer multiples of some basic energy. But for Planck, it was just a mathematical trick, waiting for a logical justification based on well-established physical properties. That is how he considered it all his life!

It was the young Einstein who, in another of his celebrated papers of 1905, had the audacity to accept Planck's mathematical conjecture as a physical reality. It is the energy of light, Einstein postulated, which is really quantized, and therefore light can be interpreted as composed of particles—quanta of pure energy. With this hypothesis, Einstein was able to explain the photoelectric effect,[3] but at the price of contradicting the wave nature of light. When physicists were finally convinced that light was a wave (*dixit* Huygens), it turned out that it was also a particle (*dixit* Newton). So much so that some years later, the particle of light was named *photon* (from *photos*: light).

The fundamental formula of quantum physics, originally proposed by Planck in his 1905 paper, is

$$E = h\nu,$$

for the energy of a photon of frequency ν, where h is what we call the *Planck constant*. It appears unavoidably in all formulas describing atomic processes. Planck's constant has the exact value $6.62607015 \times 10^{-34}$ J × seconds, by definition as it was decided in 2019. As a consequence, the value of the standard kilogram of mass has been fixed universally and no longer depends on a platinum bar in the Sèvres laboratory of France.

Thus, the old controversy about the nature of light—wave or particle?—which seemed settled, resurfaced with new impetus. Einstein had shown that the quantum of energy can be interpreted as a perfectly physical entity and no longer as a mathematical trick. This apparent paradox led to a fundamental principle of the atomic world: the duality between wave and particle, clearly and explicitly proposed by Louis de Broglie (1892–1987) in 1924. According to this principle, a particle of the atomic world, such as a photon or an electron, behaves as a wave or a particle depending on

---

[2] See Footnote 13 in Chap. 3.

[3] According to the photoelectric effect, light of a frequency above a certain critical value generates an electric current in some metals. If the frequency of the light is less than the critical one, the effect does not occur no matter how intense the light is (contrary to what would be expected if light were a wave). The theoretical explanation of this phenomenon earned Einstein the 1921 Nobel Prize in Physics.

the experimental way it is observed. Quite generally, any atomic particle with mass m and speed v can also behave as a wave with wavelength $\lambda = h/mv$, a formula where the inevitable Planck constant h appears. Indeed, subsequent experiments showed that electrons exhibit the phenomenon of interference, just like waves (see below).

Another fundamental step at the beginning of the twentieth century was the discovery of the structure of the atom. In a famous experiment designed by Ernest Rutherford (1871–1937) and performed in 1909 by his collaborators Geiger and Marsden,[4] it was shown that the atom consists of a positively charged central nucleus, compact and well localized, with negatively charged electrons all around it. The chemical properties of any element depend exclusively on the electrical charge of the nucleus and the number of electrons around it... as Faraday had visioned.

5.2 Atomic Cabala

According to a famous anecdote, Isaac Newton discovered that sunlight, passing through a prism, is split into the colors of the rainbow, namely a superposition of light of various wavelengths. A next important step in the experimental study of light occurred at the beginning of the nineteenth century, when Joseph von Fraunhofer (1787–1826) succeeded in decomposing the sunlight and "stretching" the rainbow much further than Newton had. He was thus able to observe more than five hundred dark lines superimposed on the colors of the rainbow.

The rainbow is actually the visible section to the eye of the electromagnetic *spectrum* (from radio waves to gamma rays; see Chap. 3). The so-called *spectral lines* discovered by Fraunhofer are not exclusive of sunlight. Soon, physicists in the nineteenth century discovered that dark lines always appear in the spectrum of light if some dilute gas is interposed in front of a light source. Likewise, if a gas is heated, it emits light with characteristic colors that correspond to very precise wavelengths. In fact, around 1860, the German scientists Gustav R. Kirchhoff (1824–1887) and Robert W. Bunsen (1811–1899) proved that the spectrum of each chemical element is a particular set of spectral lines that correspond to precise wavelengths (or frequencies). Something like the "fingerprints" of each element! It was a crucial discovery since it allowed chemical elements to be identified by their spectra. It was possible to determine what stars are made of without having to reach them to collect samples.

Hydrogen, in particular, turned out to have a very simple spectrum. In the section of the spectrum that corresponds to visible light, four lines appear with wavelengths 656, 486, 434 and 410 nm.[5] In 1885, the Swiss mathematician Johann Balmer (1825–1898) discovered that these numbers could be derived from a simple formula: the wavelength of each line was proportional to the inverse of.

---

[4] Hans Geiger(1882–1945) and Ernest Marsden (1889–1970).

[5] The wavelengths of visible light are limited between 380 (violet) and 750 (red) nanometers.

1/4 − 1/9,
1/4 − 1/16,
1/4 − 1/25,
1/4 − 1/36,

that is, 1/4 *minus* 1/n^2, where n is an integer from 3 to 6.

The Balmer series is typical of the visible portion of the electromagnetic spectrum. When ultraviolet and infrared detectors were developed, other series of lines, similar to Balmer's, were discovered. Finally, the Swedish physicist Johannes Rydberg (1854–1919) found, empirically, a more general formula than Balmer's with which all the known spectral lines of hydrogen could be located. The formula is relatively simple: each line corresponds to a wavelength λ given by

$$\frac{1}{\lambda} = R\left(\frac{1}{n_1^2} - \frac{1}{n_2^2}\right),$$

where n_1 and n_2 are integers (n_1 less than n_2) and R is the so-called Rydberg constant with units of inverse length; its value is 10,973,731 m^{-1}.

With $n_1 = 1$ we have the wavelengths in the ultraviolet, with $n_1 = 2$ the Balmer series is recovered, and with $n_1 = 3$ we have a series in the infrared. Spectroscopy was beginning to look like cabala, at least for the element hydrogen!

5.3 The Bohr Atom

To give some physical support to what seemed to be a game of numbers, Niels Bohr (1885–1962) proposed, in 1911, a heuristic model for the hydrogen atom—the simplest atom, with only one electron—based on the quantization of energy and the newly discovered structure of the atom. According to Bohr, the atom would be like a miniature Solar System, but unlike the planets that can be in any orbit, the electrons could only stay in well-defined orbits, corresponding to quantized energies. The energy of the orbit number n of the hydrogen atom would be $E = - E_0/n^2$, where n is an integer number and E_0 is the energy of the first and most basic orbit (The energy is negative because the electron is "bound" to the nucleus, and positive energy is needed to free it). By "jumping" from one orbit to another, the electron emits or absorbs a particle of light with exactly the difference in energy between the two orbits. This would correspond to the emission or absorption of light with a well-defined wavelength, which would be observed as a spectral line.

Bohr's model explained Rydberg's formula for the spectrum of light emitted by hydrogen atoms, but it flagrantly contradicted the known laws of electrodynamics. Indeed, according to Maxwell's theory, a charged particle in accelerated motion, as it would be in a circular orbit, should emit energy, so that, after a certain time, an electron would end up falling into the nucleus.[6] It was evident that classical physics did not apply to the atomic world. A new theory was needed.

5.4 Quantum Theory

The mathematical construction of quantum mechanics began to take shape in 1925 with the works of Max Born, Jordan, Heisenberg, Schrödinger, and Pauli.[7] The young Heisenberg, influenced by the positivism that was in vogue at the time, worked out a mathematical formulation of atomic physics based exclusively on quantities that are observable, at least in principle. Thus, in his first article of 1925, he sought to make use of only measurable quantities, such as the wavelength of light emitted by an atom, since it would make no sense to ask, for instance, what the speed of an electron is if it cannot be observed.[8]

That first article by Heisenberg lacks a solid physical foundation, but it turned out to be correct in the long run. The basic idea was to express physical quantities such as the frequency and amplitude of the emitted radiation (which can be observed) in their classical version, and "discretize" them, that is, to assume that they can be functions of only integer numbers. After all, that is what Bohr did with his model of the atom, when he postulated that the energy of electrons is proportional to $1/n^2$, where n is an integer. The sequence of integer numbers thus took the place of the continuum of real numbers, and the partial differential equations became finite difference equations.[9]

One should be aware, however, with what is meant by the multiplication of two variables in this scheme. That was clarified by Max Born and Pascual Jordan in another classical article of 1925, in which they showed that Heisenberg's idea was equivalent to interpreting physical variables, such as position and momentum, as *matrices* and not as simple numbers. Matrices, in mathematics, are arrays of numbers which have their own addition and multiplication rules (mathematics is rich enough to provide any necessary tools to physicists). Following their proposal, Born and Jordan showed that Hamilton's well-established formalism of classical mechanics could be applied directly to the description of the atomic world. It is only necessary to take quantities such as position, momentum, and energy as matrices! Thus, "matrix mechanics" was born.

---

[6] Planets in orbit around a star emit gravitational radiation, but the energy loss is totally negligible; It is only noticeable in extremely compact bodies, such as neutron stars. This is not the case for electromagnetic radiation, which is 40 orders of magnitude more energetic than gravitational radiation.

[7] Max Born (1882–1970), Pascual Jordan (1902–1980), Werner Heisenberg (1901–1976), Erwin Schrödinger (1887–1961) and Wolfgang Pauli (1900–1958).

[8] Years later, Heisenberg recalled a conversation he had with Einstein who said something that would shock any positivist: "It is the theory which decides what we can observe". Heisenberg (1972).

[9] For a function $F(x)$, its derivative in finite differences would be $[F(x + (n + 1)a) - F(x + na)]/a$, where n is an integer and a the interval considered.

In their subsequent article in 1925, Born, Jordan and also Heisenberg presented the most general version of their formalism. Among other results, they obtained the famous Planck function for blackbody radiation. Furthermore, the three authors showed with their formalism that not only energy, but also *angular momentum* is quantized. This result was to be crucial to the mathematical development of quantum mechanics.

Angular momentum, in classical physics,[10] is defined as the triple product of the mass of a particle, the radius of the orbit and its orbital speed. It has the important property of being conserved (for instance, if the speed of a satellite decreases, the radius of its orbit increases, and vice versa). In classical mechanics, an angular momentum can take any value, but in quantum mechanics it can only have quantized values, proportional to the basic unit of angular momentum... which turns out to be Planck's constant divided by 2π, or what physicists currently designate with the symbol $\hbar$ (read h bar).[11] Furthermore, angular momentum, being a vector, has an orientation in space with respect to a given axis. In that same paper, the three authors showed that the orientation of the angular momentum is also quantized. Thus, two new quantum numbers appeared.

The theoretical formulation of Born, Heisenberg and Jordan was a bridge between classical mechanics and quantum mechanics,[12] both being based on the classical formalism of Hamilton. In classical mechanics, if the Hamiltonian function is known for a given physical system (for instance, a particle subjected to an electric force), the Hamilton equations describe the evolution in time of that physical system. Similarly, in matrix mechanics, one can descend from the classical to the atomic world by "simply" taking the basic formulas of classical mechanics and rewriting the position, the momentum, and the Hamiltonian as matrices instead of simple functions. The whole procedure has no justification based on first principles... but it does work perfectly!

5.5 Uncertainty

In the atomic world, it is not the same to measure the speed and then the position of an electron than the other way around, since one measurement affects the other. Therefore, the results of two successive measurements do depend on the order of the factors. More specifically, as Heisenberg clearly argued in another classic article,

[10] The angular momentum vector is defined as $L = m\, r \times v$ (m: mass; r: position vector; v: velocity).

[11] It is a curious fact that in all important formulas of quantum mechanics, the Planck constant appears divided by 2π. The notation $\hbar = h/2\pi$ is due to Paul Dirac (whom we will meet a little later), who was probably tired of writing $h/2\pi$ each time.

[12] Heisenberg received the Nobel Prize in Physics in 1932 for his contributions to quantum mechanics. Born received it in 1954 for the same reason. Jordan never received it because he had joined the Nazi party in the 1930s.

if we decide to measure the position of an electron with great precision, we will necessarily have to alter its speed, and this would be at the cost of losing the precision with which the speed is determined. Similarly, an experiment designed to determine the speed of an electron with great precision would affect its position and make it impossible to know exactly where it is. This is the manifestation of the famous Heisenberg's *uncertainty principle*. Mathematically, this principle has the form $\Delta x \, \Delta p > h$, where Δx is the size of the uncertainty in the position x, Δp is the size of the uncertainty in the momentum p (mass $\times$ velocity), and h is the unavoidable Planck constant that appears in all the processes of the atomic world: the product $\Delta x \times \Delta p$ must always be greater than h.

This uncertainty principle applies to position as measured in the same direction, but it does not apply if the position is measured in one direction and the momentum in another perpendicular to it.[13] In the language of quantum mechanics, a pair of observables can be measured with unlimited precision if the operators that represent them do commute: in this case, it is possible to carry out joint measurements with unlimited precision. It is only a matter of choosing the most appropriate observables, those that do not interfere between themselves.

Furthermore, by having the freedom of choosing the order of the observations, the observer influences what he observes. Therefore, in quantum physics, it is no longer possible to separate the observer from the result of his observations. This fact, which is obvious if one thinks about it, has unfortunately given rise to esoteric interpretations... but we will not dwell on that.

There is also an uncertainty principle for the energy and the time. Suppose we measure the energy that a physical system possesses during a certain time Δt. The energy is measured with an error of the size ΔE and it follows that there must be a minimum uncertainty of these two joint measurements, which can be expressed with the formula $\Delta E \, \Delta t > h$, where Δt is the size of the uncertainty in the measurement of the time. In other words, the lengthier the time taken to perform a measurement, the more accurate is the measured energy, but at the expense of the precision in the time. At the other extreme, a massive particle—with energy given by $E = mc^2$—can appear and disappear in too short a time to be detected. In other words, a particle of mass m can be created *ex-nihilo* and disappear in a time shorter than h/mc^2 without violating the conservation of mass-energy! This curious effect is perfectly real and implies that vacuum in quantum mechanics is not actually empty but filled with "quantum fluctuations". As surprising as it may appear, these fluctuations, which seem to be a purely mathematical artifact, have effects that can be measured experimentally, as we will see in the chapter on relativistic quantum mechanics.

---

[13] If the particle is located at the point (x, y, z) with momentum (p_x, p_x, p_x) in cartesian coordinates, the uncertainty principle applies to pairs $\{x, p_x\}$, $\{y, p_y\}$, $\{z, p_z\}$, but not to "crossed" pairs such as $\{x, p_y\}$, etc.

5.6 The Spin

To complete the discoveries in the atomic realm, a classic experiment was carried out in 1922 by Otto Stern (1888–1969) and Walter Gerlach (1889–1979) that revealed an unexpected property of the electron, a property as fundamental as its mass and its charge. It is what physicists call the *spin*. However, despite the name, it is not about the electron spinning like a top since it is not possible to "see" something like that. It rather corresponds to the electron being like a small magnet, which can be oriented by an external magnetic field. But unlike a macroscopic magnet, which can be oriented in any direction, the electron's spin can only be oriented in two (!) directions with respect to a magnetic field. The direction of the spin is also quantized.

The spin, as a new parameter in quantum mechanics, can only be conceived by means of mathematics. We now know that all the elementary particles that make up matter have a spin which is quantized. In comparison with the angular momentum, which is a multiple of $\hbar$, the spin of the electron is *one half* that fundamental quantum constant: $(1/2)\,\hbar$. This means that an electron in an atom, besides energy and angular momentum, has also the orientation of its spin as an additional parameter… which can have only two values: spin "up" and spin "down".

5.7 Quantum Numbers

In quantum mechanics, energy is quantized, as well as the magnitude and the orientation of angular momentum. Accordingly, the state of a quantum particle is characterized by three numbers, to which the spin must be added for a total of four quantum numbers. More precisely, the magnitude of the angular momentum takes the values $\sqrt{l(l+1)}\,\hbar$, where $l = 0, 1, 2, 3$, etc. is a positive integer. Furthermore, the angular momentum being a vector, its projection onto a particular axis takes the quantized values $m_s\,\hbar$, where m_s is a positive or negative integer that runs from $-l$ to $+l$, that is: $m_s = -l, -l+1 \ldots l-1, l$. As for the electron spin, it has the magnitude $1/2\,\hbar$, and its projection along any axis can only have one of the two values $+1/2\,\hbar$ and $-1/2\,\hbar$.

Thus, all the possible states of an electron in an atom can be characterized with four numbers, each with certain restrictions, and the fact that the energy levels are quantized has fundamental consequences. There was only one problem: how can an electron in an atom avoid falling to the lowest energy level, as it should by classical mechanics? The answer was found by Wolfgang Pauli who discovered a totally unexpected property of Nature: the *exclusion principle*, according to which, in each atomic state defined by its four quantum numbers, there can be one and only one electron. Thus, each energy level has a certain number of possible states, which can be partially or fully occupied by electrons. It was the old concept of the impenetrability of matter coming true in a precise mathematical formulation.

With this relatively simple scheme, the chemical properties of the elements could be correctly explained. The fact that the lowest energy levels and angular moment states are saturated is equivalent to closing a shell; the next shell contains a certain number of electrons that can be shared with other atoms in similar conditions to form stable molecules, just as in a Lego set. The so-called noble gases are those whose levels are saturated without leaving extra electrons, so they cannot combine with other atoms. This is, of course, only a qualitative description; a more complete description must consider the repulsive interactions between all the electrons of the atom, which are not negligible, a matter that concerns the modern field of quantum chemistry. But this requires the Schrödinger equation, as we will see below.

5.8 The Wave Function

Shortly after Born, Heisenberg, and Jordan's articles appeared, their colleague Pauli used matrix mechanics to solve the important problem of the hydrogen atom. He succeeded in showing that the energy levels of the atom are quantized, just as predicted by Bohr's atomic model. Unfortunately for Pauli, his work went unnoticed[14] because, almost simultaneously in 1926, a famous article by Erwin Schrödinger appeared that would revolutionize the study of atomic phenomena.

With a stroke of inspiration impossible to explain, Schrödinger showed that the problems of quantum mechanics can be reduced to solving a partial differential equation. His basic idea was to use the wave-particle duality and wonder how a particle, say, an electron would appear as a wave: the "electron wave" should have a wavelength $\lambda = h/p$ (according to de Broglie) and a vibrational frequency $v = h/E$ (according to Planck)—here p is momentum (mass × velocity), and E is energy. In this way, Schrödinger obtained a partial differential equation somewhat resembling a wave equation and given (once again!) in terms of the classical Hamiltonian. The solution of this equation is a complex function (with real and imaginary parts[15]) of position and time. Once that solution is known, all the information about an atomic system can be deduced from it. That abstract solution of the Schrödinger equation would come to be known as the *wave function* ψ (the Greek letter *psi*).

Physical quantities such as position, velocity, or momentum are numbers in classical mechanics, and they are replaced by matrices in the Born, Heisenberg and Jordan formalism. These same variables, in Schrödinger's formalism, are replaced by what is known in mathematics as *operators*, which are rules for taking partial derivatives. For instance, the moment p_x in the x direction is replaced, in Schrödinger's formulation, by the partial derivative $\partial/\partial x$ multiplied by the factor $h/2\pi i$, where i is our well-known imaginary root $\sqrt{-1}$ and h is the Planck constant. Of course, such

---

[14] It is a curious fact that, to date, hardly any professional physicist has heard of Pauli's article: "Über das Wasserstoffspektrum vom Standpunkt der neuen Quantenmechanik", Z. Physik 36: 336 (1926).

[15] See Chap. 1, *Euler*.

an operator makes sense if applied to a function, which is, in this case, the wave function. The procedure proposed by Schrödinger consists of taking the Hamilton function of position and momentum from classical mechanics, to substitute positions and momenta by their respective operators, and apply everything to the wave function ψ. There is absolutely no rigorous justification for the formalism, but it works superbly!

In his first paper, Schrödinger applied his method to the problem of the hydrogen atom. He took the same classical formula for the energy of the electron—the simple formula resulting from Coulomb's law—[16] and showed that his newly presented equation admits physical solutions in terms of certain functions—known as Laguerre[17] polynomials—if the energy is quantized... according to the Bohr formula! Only solutions that satisfy this condition are physically acceptable because they correspond to bounded solutions for the wave function.

But what kind of wave is the wave function ψ? Does it represent something physical or is it just a mathematical construct? Schrödinger first speculated that it would represent the electric charge density of the electron, so that this particle, rather than being an indivisible point, would be more like a fluid. Accordingly, he initially thought that it was necessary to return to the concept of matter as a continuum. However, Max Born proposed a different interpretation that turned out to be perfectly consistent. According to Born, the squared modulus[18] of the wave function, which is $|\psi|^2 = \psi \, \psi^*$, would represent the probability of finding a particle with certain properties, such as energy or position; for instance, $|\psi(x, t)|$ would be the probability of finding the particle at position x at time t. Schrödinger never liked this interpretation, but it is the one that became fully accepted and survives to this day... with complete success!

Regardless of the interpretations, the works of Born, Heisenberg and Jordan, on the one hand, and Schrödinger on the other, established not only the foundations of quantum physics, but also those of modern chemistry, since they explained the electronic configurations of the atoms and the ways they combine with each other to form molecules. The Schrödinger equation has been, since its formulation, the basis of most investigations in quantum physics, since it allows us to solve the problems of the atomic world with great precision, notwithstanding how complicated they may be. It is the fundamental basis of quantum chemistry. However, its derivation appears due to a magical act of intuition and remains as obscure as it was formerly. Schrödinger's formulation did not supersede matrix mechanics but, on the contrary, justified and complemented it, since it was later shown that the two formalisms are equivalent and yield the same results. It is only for each particular problem that it is more advantageous to use one or the other.

---

[16] The formula is $E = Qq/r$ for the energy E of a particle with charge q in the field of another particle of charge Q at a distance r. In this case, the Hamiltonian is $H(r,p) = p^2/2\,m + E$.

[17] For Edmond Laguerre (1834–1886), French mathematician.

[18] A complex number z has the form $z = x + iy$, where x and y are real numbers. The conjugate of z is defined as $z^* = x\text{-}iy$ and the squared modulus of z is $|z|^2 = zz^* = x^2 + y^2$, which is a real number.

5.9 Copenhagen

The new physics of the atomic world was based on solid mathematical concepts, but their physical meaning was a mystery. An explanation had to be found, even though in new and counterintuitive terms. It is for this purpose that Niels Bohr proposed the "Copenhagen interpretation", which was quickly accepted by a new generation of physicists who were not afraid to confront the canons of classical rationalism defended by their elders, such as Planck, Schrödinger, and Einstein.

Bohr sustained that, with quantum mechanics, we must be satisfied with a description in purely mathematical terms and give up the pretense of understanding the atomic world with concepts of our macroscopic world. For instance, the spectral lines of atoms are produced by "quantum jumps" of the electrons from one orbit to another, with the consequent emission or absorption of photons. The probability of a particular jump to occur can be calculated from the Schrödinger equation, but it is a spontaneous jump during which the particle loses its "physical reality" and the equation cannot describe it during the "jump". In fact, quantum mechanics cannot calculate when a quantum jump will occur, but only the probability that it will occur in a certain time.

As for the notorious wave function which is the solution of the Schrödinger equation, the Copenhagen interpretation fully accepted Max Born's proposition that it was a probability wave which represents the states of a physical system. A classical experiment, related to this interpretation, consists of sending a beam of electrons through two close openings: the electrons arrive one by one on a screen, apparently randomly at first, but, after a while, the characteristic interference pattern begins to appear (a succession of dark and bright stripes).[19] This pattern is not due to an electron interacting with another one, but to the interference of the wave function of the *same* electron with itself.

Regarding the interference between the observer who performs an experiment and the result of his observation, the situation can be summarized as follows: if an experiment is designed to see electrons as waves, they behave as waves, and if designed to see them as particles, they behave as particles. What happens actually is not that the electrons guess the intentions of the experimenter, but that the experiment influences the result. As Richard Feynman pointed it out,[20] herein lies the fundamental mystery of quantum mechanics.[21]

The wave function is a probability wave, but the concepts of "classical" and "quantum" probabilities are different. Quantum states can interfere with each other through their probability waves. This property is well represented in mathematical language by the algebraic properties of complex quantities. From a mathematical point of view, the wave function should be something like the "complex square root" of classical probability. For instance, if an atomic system admits two states,

---

[19] An experiment first carried out convincingly in 1989 by Tonomura et al.

[20] Richard Feynman (1918–1988). American physicist, probably the last of the most influential physicists of the twentieth century.

[21] Feynman (1968) *Lecture notes in physics (Vol. III)*. See also Feynman (1995) *Six Easy Pieces*.

represented by the wavefunctions ψ_1 and ψ_2, the wavefunction of the joint system is $\psi = \psi_1 + \psi_2$, and the probability of the state of system 1 or system 2 is $|\psi_1|^2$ and $|\psi_2|^2$, respectively. However, the total probability of finding the system in one state or the other is[22]

$$|\psi|^2 = |\psi_1|^2 + |\psi_2|^2 + \psi_1 \psi_2^* + \psi_2 \psi_1^*.$$

The first two terms represent the usual individual probabilities, and their sum is just the classical probability for the occurrence of both states; *however*, the last two terms correspond to the interference between the two different states, something with no counterpart in the macroscopic world. Thus, wavefunctions can interfere with each other...just like real waves in water! The difference between the atomic world and the macroscopic world lies precisely in this kind of interference, which vanishes in the macroscopic world. The loss of coherence between the different quantum states, called *decoherence*, is what produces the emergence of the macroscopic world from the quantum world. In practice, we do not notice decoherence because it takes place in a typical atomic time scale.[23]

Nonetheless, Born's probabilistic interpretation was not without conceptual difficulties. The Schrödinger equation describes the evolution of the wave function in a deterministic way, which can be calculated exactly from its initial value. Accordingly, quantum mechanics, described by this equation, should be perfectly deterministic. However, the wave function is a superposition of all possible states, and the observation reduces this set of states to a single one: the state that becomes the observed one. It is said that the wave function "collapses", and this collapse is not deterministic; one can only calculate the probability that it will occur.

The polarization of light provides a good example of the superposition of states with the consequent collapse of its wave function. The polarization is determined by the direction in which light vibrates as an electromagnetic wave, and a polarizing filter has the property of letting through only the fraction of the light that vibrates in a single direction. Accordingly, if light is made of photons, only a certain fraction of them get through a polarizing filter... but what about a single photon? Since it cannot be divided, all we can say is that it did went or did not through the filter. In this case, the interpretation given by quantum mechanics is that, before reaching the filter, the wavefunction of the photon is a superposition of two wavefunctions corresponding to two possible states: a state in which the photon will go through and a state in which it will not. It is upon reaching the filter that the double wave function of the photon collapses into a single one.

This interpretation of a photon polarization, and many more like it in quantum mechanics, are perfectly consistent, but they cast serious philosophical doubts on what it is meant by physical reality before it is observed. Presently, we can only argue that the Copenhagen interpretation leads to theoretical results that can be

---

[22] See Footnote 16 in Chap. 4.

[23] The typical size of an atom is of the order of 10^{-10} m and the time taken by light to traverse this distance is about 10^{-18} s, a typical atomic time.

verified experimentally, even if they seem paradoxical (see the experimental scheme proposed by Bell in the next Chapter).

5.10   The Spin (2)

The spin turned out to be a fundamental parameter characterizing all subatomic particles, a parameter as important as mass and electric charge. It permits to classify all particles into two large groups: fermions and bosons, named after Enrico Fermi (1901–1954) and Satyendranath Bose (1894–1974) respectively, who described their statistical properties. Fermions are the particles of matter and are distinguished by the fact that their spin value is a half-integer of $\hbar$ (1/2 for electrons, protons, and neutrons or 3/2 for exotic particles), while the spin of the bosons is an integer multiple of $\hbar$ (1 for the photon, or 2 for the hypothetical graviton); bosons produce the interactions between the particles.

Bose's contribution was to show that Planck's famous formula for the blackbody radiation can also be obtained from purely statistical arguments, just by counting how many ways a set of particles can be distributed in a given number of cells. The only additional assumption that must be made is that the particles are indistinguishable from one another, which is a peculiarity of the particles of the atomic world, and particularly of photons. Bose's paper was rejected for publication by his English colleagues, who did not believe that Planck's formula, so well established on physical grounds, had such a purely statistical interpretation. Bose then sent his article to Einstein to ask for his opinion; not only Einstein did understand the importance of the work but he himself translated it into German for its publication.

Bose's statistical method can also be applied to particles that satisfy Pauli's exclusion principle. Taking this into account, Fermi and, independently, Dirac[24] followed Bose's idea to deduce a formula equivalent to Planck's, but for particles that obey Pauli´s principle, that is, particles that are indistinguishable within them and, furthermore, only one particle fits in one quantum state.

Unlike fermions, bosons do not satisfy the exclusion principle, so that any number of bosons can be condensed into the same minimum energy state. This can happen at temperatures very close to absolute zero: it is the "Bose–Einstein condensation", theoretically predicted by Einstein after reading Bose's article. It can be called a new state of matter, which was produced in a laboratory for the first time in 1995.[25] Natural philosophers of past centuries, who considered the impenetrability of matter to be one of its primordial properties, obviously did not know about bosons.

---

[24] Enrico Fermi (1901–1954), Italian-American physicist, 1938 Nobel Prize for his important contribution to nuclear physics. Paul Adrian Maurice Dirac (1902–1984), British physicist, 1933 Nobel Prize. We will meet him many times in the following Chapters.

[25] By Eric Cornell, Carl Wieman and Wolfgang Ketterle, who received the 2001 Nobel Prize.

5.11 Least Action (Again)

In classical mechanics, the Lagrangian and Hamiltonian functions are directly related to an action which is defined mathematically in terms of functions of the positions and velocities (or momenta) of a physical system subjected to various forces. The action turns out to have its minimum value if the positions and velocities of the system satisfy the physically correct equations of motion. The action must be invariant, in the sense that it does not depend on the coordinates or the reference system used, as required by the principle of relativity.

As for quantum mechanics, it can also be formulated in terms of Lagrangians and Hamiltonians as functions of operators representing positions and momenta. The principle of least action or Hamilton principle (see previous chapter) holds: the action is minimal if the physical system satisfies the Heisenberg or Schrödinger equations of quantum mechanics. If quantum mechanics were to be defined by a single sentence, it could be said that it is "the theory in which the action is quantized, and the quantum of action is equal to the Planck constant". (Note that action and Planck's constant have the same physical units: energy × time, or mass × velocity × distance.)

Additionally, the old concept of force disappeared definitively in quantum mechanics to leave the place to such abstract concepts as "wave function", its "collapse", "quantum jumps", "superposition of states", "quantized fields", "quantum fluctuations of vacuum", and so on. The concepts of energy and field have survived, fully and vigorously, together with the appropriately adapted formalisms of Lagrange and Hamilton, as well as the principle of least action.

The mathematical structure of quantum mechanics was rigorously established by John von Neumann (1903–1957) in his influential 1932 book *Mathematical Foundations of Quantum Mechanics*, in which he proved, in particular, that the basis of quantum mechanics is a mathematically defined space: the Hilbert space (another contribution of Hilbert to mathematical physics!). Basically, a Hilbert space is a wide-ranging generalization of ordinary space, which can have any number of dimensions including infinity, and in which points are defined by complex numbers and there is a definition of distance between points. Von Neuman showed that the wave function of quantum mechanics can be interpreted as a vector in Hilbert space, with its magnitude being the (squared) probability.

Thus, the mathematics that underlies quantum mechanics started to emerge together with its physical basis. Every physically meaningful concept turned out to have its corresponding mathematical concept. For instance, the uncertainty principle has a precise mathematical representation in terms of matrices or operators, since these mathematical objects do not commute: AB is not the same as BA, as the order of two measurements yield different result. To complete the correspondence between classical and quantum formalism, Dirac showed that the so-called commutator, AB minus BA, is in the Born, Heisenberg and Jordan formalism the exact equivalent of what is known as *Poisson parenthesis* in classical mechanics,[26] a concept that

---

[26] After Siméon Denis Poisson (1781–1840), French physicist and mathematician.

appears in mechanics *à la* Hamilton which relates two physical quantities such as position and momentum. It seems, then, that there is a deep mathematical structure that occurs in various ways, depending on the mathematical language being used.

Alternative formalisms to those of Heisenberg and Schrödinger are still possible, such as Feynman's idea of using the principle of least action to define the "path integral", in which the different possible histories of a process are "added up". The essential idea is that all trajectories are possible in principle, but they produce interferences among them that cancel each other, except the real trajectory that minimizes the action and remains. It is a perfectly valid and ingenious formulation that has enlightened some problems of quantum mechanics, but it is rather limited in its applications.

Chapter 6
Quantum Paradoxes

Abstract The famous Schrödinger's cat paradox and its variant, the Wigner's friend paradox, are discussed. These paradoxes are related to the phenomenon of decoherence which regulates the passage from the quantum to the microscopic world. The Einstein, Podolsky and Rosen paradox reveals the non-locality of the quantum world. It is noteworthy that non-locality can be proved experimentally.

The intention to explain the processes of the quantum world by means of words of the common language inevitably leads to contradictions. This fact was noticed since the early days of quantum mechanics, when the "elder" physicists were still looking for an interpretation of quantum mechanics in terms of "reasonable" concepts, instead of the Copenhagen interpretation, which leads to several apparent paradoxes.

6.1 Cats and Friends

The act of observing is what assigns reality to things, as Bohr maintained in his philosophical writings.[1] In this regard, an apparent –and very famous– paradox is that of the "Schrödinger's cat", proposed by the author of the famous equation. The idea is to relate a macroscopic process with an atomic one, such as the decay of a radioactive nucleus. Suppose, Schrödinger proposed, that a cat is confined in a box, inside which there is a radioactive atom, along with a mechanism linked to a Geiger counter that breaks a bottle of poisonous gas when radiation is detected. According to the Copenhagen interpretation, the radioactive atom is in a superposition of two states: in one state it has emitted radiation, in the other it has not yet. Therefore, if the box is not opened to observe its content, the cat would be in a superposition of two states... alive and dead!

---

[1] N. Bohr, *Philosophical Writings*, Volumes I, II, III (Ox Bow Press).

© The Author(s), under exclusive license to Springer Nature Switzerland AG 2023
S. Hacyan, *The Mathematical Representation of Physical Reality*,
The Frontiers Collection, https://doi.org/10.1007/978-3-031-21254-3_6

A variant of the Schrödinger's cat paradox is the "Wigner's friend" paradox.[2] with which the absurdity of the situation is even more evident. As Eugene Wigner[3] proposed, suppose that instead of a cat we put a human being in the box (obviously without a killing mechanism, but something more benign, say, a lightbulb that goes on or off). That human being can calculate the probability that the atom emits radiation and observe when it does. However, to another observer outside the box, the atom and his friend inside, are in a superposition of two states, with quantum interference between them. Just when the box is opened, the joint wave function of the atom and the friend collapse, without the latter obviously feeling anything strange.

Once again, the solution to the paradoxes—the cat and Wigner's friend—can be sought in the decoherence phenomenon we mentioned in the previous chapter: the superposition of states applies to atomic systems, but it is completely imperceptible and irrelevant for systems made up of billions and trillions of atoms because the interferences (quantum effect!) between so many states are cancelled in a very short time, a typical time of atomic processes.[4] The interesting point is that these effects have been verified with single atoms, where interference and decoherence cannot be neglected.[5]

6.2 Non-locality

In quantum mechanics, causality is intimately related to locality. Causality is under-stood as the fact that, since the speed of light is the maximum speed allowed in Nature, an event E can influence, at least in principle, only those future events that can be reached from E at the speed of light. Locality, in turn, refers to the fact that a region can only be influenced by events causally related to it. In other words, there is no instantaneous communication in Nature.

In a famous work of 1936, Einstein, Podolsky and Rosen (EPR)[6] clearly demon-strated that the Copenhagen interpretation is incompatible with the principle of causality and locality, and, consequently, with what they called "local realism". In the words of EPR: "If, without perturbing a system, we can predict with absolute certainty the value of a physical quantity, then there is an element of physical reality that corresponds to that physical quantity." In other words, the physical properties that can be observed, even if only in principle, exist independently of being observed, while the Copenhagen interpretation asserts that the physical properties of an atomic system acquire physical reality only when they are observed (i.e., mathematically

---

[2] Wigner (1961).

[3] Eugene P. Wigner (1902–1995), Hungarian born physicist. Nobel Prize in 1963 for his many contributions to mathematical physics.

[4] Hacyan (2015).

[5] The 2012 Nobel Prize in Physics went to Serge Haroche and David Wineland for their experiments with atoms and photons that confirmed the strange effects of quantum mechanics.

[6] Boris Y. Podolsky (1896–1966). Nathan Rosen (1909–1995).

speaking, when the wave function collapses). The purpose of EPR was to pinpoint the paradoxical situation that two quantum particles, with a common origin, can form a single quantum state—that is, share a single wave function—[7] and that, therefore, the measurement made on one of the particles affects and defines the "physical reality" of the other, even if it is kilometers or light years away. This would violate the principle of causality, so fundamental for the theory of relativity. In this regard, Einstein once declared that this would imply the existence of a "spooky (*spukhafte*) action at a distance".[8]

For EPR, the paradox was the proof that there are "elements of physical reality" that exist regardless of whether they are observed or not. Therefore, either causality and locality do not apply in the atomic world, or quantum mechanics is an incomplete theory because it cannot deal with all the elements of physical reality. For EPR, this second option was the correct one. It meant that there were properties of the atomic world that could be described, in a better theory, by "hidden variables"... hidden from quantum mechanics but real.

Einstein never questioned the validity and efficacy of quantum mechanics, but he always thought that there should be a deeper theory to support it, a theory that could describe the variables that were still hidden. There were various attempts to find this supposedly deeper physical theory, but without success. Finally, in the 1960s, John S. Bell (1928–1990), using a mathematical argument as simple as ingenious, showed that it is possible to verify experimentally whether hidden variables exist; in other words, if quantum mechanics is the deepest theory we can build, despite violating locality and the principle of causality.

The scheme proposed by Bell is well illustrated by the polarization of light. As we saw in the previous Chapter, a photon has two possible directions of polarization that correspond to whether it will get through a polarizer filter or not. Suppose two photons are emitted from a common origin in two opposite directions and both are detected through two polarizer filters. One can count the number of photons that go through each filter, set at random orientations.[9] Following the original idea of EPR, Bell showed that the numbers of photons observed at each position are strongly correlated between them, more strongly than would be expected with a classical scheme based on hidden variables (the correlation is a measure of how much one set of data affects another). The crux of the matter is that the spatial separation of the two photons can be arbitrarily large, as we have pointed out before.

If this matter had to be explained with words, the only explanation would be that the photons on one side "guess" instantly what happens to their partners on the other side and "decide", accordingly, how to go through their filter. In Bell's experimental proposal, it is simply a matter of counting the number of photons going through each of the two filters, set at various angles, and then calculate the correlations between the data of both measurements. According to the prediction of quantum mechanics,

---

[7] An "entangled state", in the terminology of quantum mechanics.

[8] Apparently in a letter to Max Born of March 3, 1947.

[9] The inclusion of the spin in the original EPR scheme for clarity is due to David Bohm.

based on simple mathematical arguments, these correlations should be greater than the classical correlations with photons behaving as classical particles.

The importance of the scheme proposed by Bell is that it is no longer the case of a *gedankenexperiment* (thought experiment) but that it can be tested in a real laboratory in the material world. Starting in the 1970s, several laboratories began to carry out experiments with photons being emitted to distant positions.[10] The obtained correlations turned out to be as predicted by quantum mechanics. The experiments showed that there were no hidden variables, but a "spooky action at a distance" instead.

Quantum correlations, which are stronger than classical correlations, are characteristic of the quantum world. Once again, because of the decoherence that we mentioned above, the "spooky action at a distance" is completely erased when emerging from the quantum world to the macroscopic world. In other words, the principle of locality, according to which information cannot be transmitted faster than light, is a statistical property of the macroscopic world, but that same principle does not apply in the atomic world.[11] In any case, quantum effects can be used to transmit part of the information in combination with traditional means, and thus reconstruct the total information. It is remarkable that a whole new branch of physics, quantum information theory, has now developed based on so paradoxical effects.[12]

---

[10] Clauser and Shimony (1978). Alain Aspect and collaborators in France: Aspect (1982).

The Nobel Prize in Physics 2022 was for A. Aspect, J. F. Clauser and A Zeilinger "for experiments with entangled photons, establishing the violation of Bell inequalities and pioneering quantum information science".

[11] In a previous text (Hacyan, 2004), I proposed that the lack of locality in quantum mechanics is an evidence that space and time are just forms of perception, as postulated by Kant (more on the subject in Part II).

[12] See, *e.g.*, Bouwmeester et al. (2000): *The Physics of Quantum Information*.

Chapter 7
Quantum Mechanics and Relativity

*Quicquid autem sit, hic calculo potius quam nostro indicia est
fidendum atque statuendum, nos saltum, si fit ex infinitio in
finitum, penitus non comprehendere.*
Leonard Euler, *Mechanica* (Chap. 3, Prop. 32, Scholium 2).
(*Be that as it may, here we must rely on established calculus
rather than on our knowledge, if the leap from infinity to finite is
not fully understood.*)

Abstract Dirac achieved the unification of quantum mechanics and special relativity. He obtained an equation which had surprising physical consequences. Infinite quantities appeared in the calculations, but Dirac found an ingenious way to deal with them, and thus predicted antimatter. The problem of infinite quantities, however, will always be present in the theory of quantized fields that describes subatomic particles.

Both the Heisenberg and Schrödinger formalisms are, in a way, quantum versions of classical mechanics, since they use the classical Hamiltonian function, but with operators in the form of matrices or partial derivatives. instead of quantities represented by numbers. The two formalisms are the quantum versions of classical mechanics and therefore do not consider relativistic effects. They do solve many problems of atomic and molecular physics, but some situations may require a treatment within the framework of the theory of relativity. Thus, shortly after the publication of Schrödinger's famous paper, physicists sought for a relativistic generalization of his equation.

7.1 Dirac Through the Mirror World

The first attempt to find the relativistic version of the Schrödinger equation is due to the physicists Oskar Klein (1894–1977) and Walter Gordon (1893–1939), who simply took the relativistic formula for the energy of a particle and, following Schrödinger's method, replaced dynamic variables, such as momentum and energy, by operators in the form of partial derivatives with respect to position and time.

The result was a second-degree partial differential equation for a "relativistic wave-function" which described a particle without spin (it could not be an electron). In any case, it seemed quite acceptable except for an annoying detail: its solutions described states with both positive and negative energies… without lower limit! As every physical system tends to fall from a state with a certain energy to another with less energy if nothing prevents it, the lack of a lower limit would imply than matter was unstable, as if it were falling into a bottomless energy pit.

In 1928, shortly after the Klein-Gordon attempt, Paul A. M. Dirac found a different and highly original solution to the problem of uniting quantum mechanics with relativity. He realized that it was more convenient to use wave-functions with four independent components instead of only one[1]—Pauli had proposed a "double" form of Schrödinger's equation to take care of the two directions of the electron spin—. Dirac managed to rewrite the formula for the energy in a very ingenious way using once again matrices. Following the already established procedure of changing energy and momenta to operators in the form of partial derivatives, he thus obtained a first-order partial-derivative equation for the four components of his wave-function (unlike the Schrödinger or Klein-Gordon equations, which are of second order for a single wave-function). Surprisingly, it turned out to be much more than a simple mathematical trick. Dirac's equation correctly took into account the spin of the electron and predicted—as a bonus!—the *magnetic moment* of the electron (basically, the strength of its intrinsic magnetic field). It was something known experimentally but not explained by Schrödinger's equation.

Nevertheless, despite its success, the Dirac equation, like the Klein–Gordon equation, contained solutions with unbound negative energy states. But the equation was too beautiful–if one can speak of mathematical beauty–to be dismissed because of that annoying detail, so Dirac himself found a rather unusual solution to the problem. For this, he resorted to the Pauli exclusion principle, according to which one and only one electron can be in one physical state. Suppose then, Dirac postulated, that each one of the infinitely many states with negative energies is occupied by a corresponding electron. That would be one possible definition of vacuum: an infinitely deep pit filled up to the top with infinitely many electrons! Then, by Pauli's exclusion principle, electrons with positive energy cannot fall to states with negative energy because all these states are occupied. Moreover, if an electron is missing in that sea of negative energy, the hole—an absence of negative charge—would manifest itself as an "electron" with positive charge; it would be an "anti-electron". Furthermore, if an electron with positive energy were to fall into that hole, the two would annihilate and, by the conservation of energy, photons would be emitted with energies at least equal to the mass-energy of two electrons Thus, based on an argument that seemed ludicrous, Dirac intuited the existence of *antimatter*. The year after the appearance of his article, the "anti-electron" or *positron*, as it is now known, was discovered in cosmic rays: it is a particle with the same properties (mass and spin) of the electron,

---

[1] In mathematical terms, it is called a *spinor*, in a certain sense something intermediate between a scalar and a vector. Hamilton's *quaternions* can be considered the ancestors of spinors.

except for its charge, which is positive. And indeed, matter and antimatter annihilate each other and produce gamma rays if they come into contact. Once again, a mathematical "trick" gave rise to a theoretical prediction that was fully confirmed in Nature.

Dirac's equation revealed the wealth of the mathematical world and its power to solve problems of the atomic world, where common sense is totally inadequate. At the same time, such a surprising treatment of infinity confirmed that physics does not always require the logical rigor that is so important to mathematicians. In fact, it was a preview of the difficulties that were to appear in the quantum treatment of the electromagnetic and other fields... and the way to solve them.

7.2 The Quantized Field

Quantum Field Theory (QFT) describes various fields as continuous functions of space and time. The field itself depends on certain physical parameters with units of length, time, and mass-energy. In this theory, as the term implies, the field itself is quantized, or rather its energy (or equivalently its mass).[2] Each quantum of energy of the field is interpreted as a quantum particle. Matter itself is considered a fluctuating field, and particles are the irreducible quantum of the energy of these fluctuations. Also, what was previously understood as a force between particles, is interpreted as an interaction mediated by particles, such as the photon (which is a quantum of electromagnetic energy).

As for vacuum, the situation is quite different at the quantum level, since space is full of "quantum fluctuations of the vacuum", as we saw in the previous chapters. These fluctuations can be interpreted as "virtual" particles of energy that appear out of nowhere and disappear in a time too short to be detected due to the Heisenberg uncertainty principle. The quantum vacuum, which is not at all empty, can be well defined mathematically and its physical effects are perfectly real and observable. The only snag is that the total energy of those quantum fluctuations, mathematically speaking, is infinite! There is no limit to virtual particles appearing and disappearing with any energy, at least in theory... but after Dirac's treatment of infinity, that did not worry theoretical physicists.

QFT had its first great successes in the 1950s, with the formulation of *quantum electrodynamics* (QED).[3] Based entirely on the relativistic version of quantum mechanics, this theory made it possible to study electromagnetic phenomena in the atomic and subatomic world with a precision never before achieved in science. For instance, it was possible to calculate the value of the "anomalous magnetic

---

[2] There have been some theoretical attempts to quantize space-time, but without clear results.

[3] The formalism is mainly due to Richard P. Feynman, Sin-Itiro Tomonaga, and Julian Schwinger, who received the 1965 Nobel Prize in Physics for their work.

moment"[4] of the electron with a precision of ten (!) significant figures, and that value was confirmed experimentally... with the same precision. The theory also explained a slight shift in the basic energy of the electron in the hydrogen atom.[5] Both effects are due precisely to the interaction of an electron with the quantum fluctuations of vacuum. Thus, the existence of these fluctuations was confirmed, despite the mathematical difficulty that their infinite energy represents.

The basic equations of quantum electrodynamics are usually presented in terms of a Lagrangian function (as expected!) involving the electrons, as proposed by Dirac, in combination with the electromagnetic field, as given by Maxwell's equations. All this leads to a series of partial differential equations that can be solved with successive approximations. The method consists of using a series expansion of the solutions (that is, a series in which each term is smaller than the previous one[6]). The procedure leads to correct and verified results, but it suffers, again, from a very annoying detail: infinite quantities appear in the resulting formulas! These are basically integrals of unbounded functions over an infinite range of their variables (a simple example: the integral $\int x\,dx$ from 0 to ∞).

How to deal with infinity? Theorists noted that the formally infinite quantities in their formulas always appeared multiplying parameters such as charge and mass. Ultimately, one can redefine those constants in such a way that they absorb the infinite quantities with which they appear. It would be like multiplying zero by infinity, which can be any number... a number that can be reinterpreted as the true charge or mass of a particle. The method, called *renormalization*, lacks mathematical rigor, but it works surprisingly well. Of course, attempts have been made to find some rigorous justification of the method, but all the efforts of mathematical physicists have been in vain. It only remains to accept that the trick works.

As for Dirac's strange "sea" of negative energies, Feynman found another interpretation: The antiparticles would behave exactly like their corresponding particles, but with time... flowing backwards! In other words, if we film the interactions between particles and then project the film backwards—and, also, watch it reflected in a mirror—we would see exactly what the corresponding antiparticles would do. The above is a mathematical result: the most general form of a Lagrangian function that can be formulated in QFT—that is, one which is invariant under Lorentz transformations (i.e., compatible with relativity)—, such Lagrangian remain invariant under the simultaneous inversion of the signs of all the charges, and the inversion of the spatial and temporal coordinates; what is technically known as CPT invariance (C: charge, P: parity, T: time)-. Any violation of CPT invariance would be due to a violation of the principle of relativity (some physicists have searched for such violation, without conclusive results so far).

---

[4] It is a slight correction to the value of the magnetic moment of the electron predicted, in a first approximation, by Dirac's equation.

[5] This is the so-called Lamb effect, calculated theoretically by W. E. Lamb (1913–2008). Nobel Prize in Physics 1955 for his work.

[6] For instance, the series $1/2 + 1/4 + 1/8 + 1/16 + 1/32$, etc. tends to the value 1 if more and more terms are taken. Each term is half the previous one.

Finally, we can point out that quantum electrodynamics could be extended, as a branch of QFT, to the nuclear forces, although the treatment is considerably more complicated, as we will see in the following.

7.3 Down the Nucleus

The discovery of the proton is attributed to Ernest Rutherford, who discovered experimentally that the nucleus of the hydrogen atom (which is a single proton) is present in all other atomic nuclei. Some years later, Chadwick discovered the existence of the neutron,[7] a particle very similar to the proton, but slightly more massive and without electric charge. Thus, the structure of the atomic nucleus began to emerge as a conglomerate of protons and neutrons. Evidently, there should be a force between the nuclear particles much stronger than the electromagnetic force, to avoid the protons to be repelled between them due to their positive charges. Furthermore, such a nuclear force should be of very short-range, basically restricted to the nucleus.

Additionally, physicists in the 1930s discovered that, beside the strong nuclear forces that give cohesion to the nucleus, there is another interaction that is responsible for the *beta decay*, the process by which some atomic nuclei emit an electron at the cost of transforming one of their neutrons into a proton and thus moving one step up in the periodic table (the inverse decay can also occur: a proton emits a positron and the atom step down one place). It became clear that it was due to another kind of force, much weaker than the nuclear one, but also restricted to nuclear distances. In brief, there were two very different nuclear forces, the strong and the weak, which had to be added to the already known forces of the macroscopic world, electromagnetism and gravitation.

7.4 Nuclear Interactions

After the great success of QED in the 1950s, quantum field theorists tried to extend that theory to nuclear interactions. As a first step, a good theory for the weak interactions was needed. It turned out that in addition to the photon that produces the well-known electromagnetic interactions, there were two other types of particles, called W and Z, which have a role like that of the photon, but with the particularity that they are extremely massive. Unlike photons, there are of very short-range due to their masses; their interactions are restricted to the scales of atomic nuclei only.[8] It was then possible to unify the weak interactions with the electromagnetic ones to form a

---

[7] James Chadwick (1891–1974), English physicist. 1935 Nobel Prize for his discovery.

[8] How can particles hundred times more massive than protons exist in the atomic nucleus? Because they appear and disappear in too short a time to be observed, just as quantum fluctuations.

single interaction: the *electroweak*. The existence of these particles was theoretically predicted and then verified, in the seventies, in large particle accelerators.[9]

In the end, quantum field theory would be extended to strong nuclear interactions, as we will see in the next Chapter. Unfortunately, it is no longer possible to perform calculations "by hand", with pencil and paper, as was the case with quantum electrodynamics. There is no other choice than to trust the results produced by computers with very sophisticated programs, and compare them with experimental results (if possible).

---

[9] The theory of electroweak interactions was formulated by Sheldon Glashow, Steven Weinberg and Abdus Salam (Nobel Prize 1979), and its renormalization was demonstrated by Martinus Veltman and Gerardus 't Hooft (Nobel Prize 1999).

Chapter 8
Other (Mathematical) Spaces

Abstract Abstract mathematical space turned out to be quite useful in the description of subatomic particles. The quark model, first predicted theoretically, has been most successful in the study of elementary particles as part of the Standard Model which encompasses electromagnetic and the strong and weak nuclear interactions. However, a quantum theory of gravity has eluded all successful attempts so far (viz., superstrings).

Mathematics achieved an exceptionally high degree of abstraction and refinement in the study of matter at the subatomic level, the world of elementary particles and their interactions. This is a level of reality that can only be described (understood?) by mathematics. For this reason, we devoted some more space, in this and the previous chapters, to the physics of elementary particles and quantum field theory.

8.1 Symmetry

To describe the nuclear force, Heisenberg remarked that the proton and the then recently discovered neutron differ only by their electric charge and slightly by mass. They would be the same particle in an imaginary world in which there were no electromagnetic interactions. Accordingly, if only nuclear forces were considered, proposed Heisenberg, we could interpret the proton and the neutron as two states of the same particle in a purely mathematical space. Moreover, it would be analogous to the electron spin, which has also two states (in the real space): up and down. The proton and the neutron would be equivalent to spin up and spin down states in this abstract mathematical space! This new characteristic of particles was given the name *isospin*. Its similarity with spin and, in general, with angular momentum (which is also quantized) turned out to be very successful, since Heisenberg was able to deduce the form nuclear interactions should have... in the additional abstract space. It would be analogous to the electromagnetic Lagrangian, which is such that it depends on the magnitude of the spin of an electron but not on its orientation in physical space (the components of a vector depend on the system of reference, but

S. Hacyan, *The Mathematical Representation of Physical Reality*,
The Frontiers Collection, https://doi.org/10.1007/978-3-031-21254-3_8

its magnitude is the same in any system). The procedure could successfully explain, at a semiquantitative level, some properties of the strong nuclear interactions, such as the close values between the masses of several nuclei with the same magnitude of isospin, or the relation between their decay rates. But the idea of using symmetrical properties in abstract spaces, additionally to physical space-time, proved to be quite successful, particularly to explain the avalanche of new particles that were being discovered day after day in the second half of the twentieth century.

8.2   Quarks

By the 1960s, physicists working with large particle accelerators had discovered an overwhelming number of new elementary particles. Besides those already known, exotic particles began to appear, with the only common characteristic of being highly unstable: they appeared and decayed in atomic time scales. These exotic particles are generated by the collision between protons or electrons that are accelerated to speeds close to the speed of light, thus transforming their energy into the mass of new particles, which disintegrate into other particles in a few millionths of a second. Faced with such a chaotic situation, physicists began to suspect that this zoo of unstable particles could be reduced to a small number of more elementary particles. The situation would be, on a deeper level, like that of the chemical elements: all the properties of the atoms listed in the Mendeleev table can be explained with the combinations of only three fundamental particles, protons, neutrons and electrons.

What would these even more elementary particles be like? The answer came from a branch of mathematics called "group theory" which deals essentially with properties of symmetry, not in common space but in abstract mathematical spaces, just like Heisenberg's original idea of interpreting the proton and the neutron as isospin states in an abstract space.

Murray Gell-Mann (1929–2019) and George Zweig in the 1960s, noticed that the states of elementary particles, both stable and unstable, could be described with the spin, the isospin and an additional variable, which they called the *strangeness*, for the lack of a better name.[1] For instance, the *pion*, a newly discovered particle, appeared with three possible electrical charges: positive, negative and neutral; which could be interpreted as the three states of the same particle with isospin equal to 1 and "projections" of isospin -1, 0 and $+1$. As Gell-Mann and Zweig noticed, it is like a rotational symmetry in an abstract space, complete with quantization just as the spin and the angular momentum are quantized in physical space. In this way they were able to classify the particles according to a few quantized parameters: charge, spin, isospin, strangeness.

Without getting into technical details, suffice it to point out that using the properties of symmetry under "rotations" in an abstract space of... eight dimensions(!), it was possible to classify the known particles into well-defined groups, and even predict the

---

[1] The physical spin has units of $\hbar$, but isospin and strangeness are taken as pure numbers.

existence of more exotic particles, which were experimentally detected and identified, thus giving a strong support to the model proposed by Gell-Mann and Zweig.

Furthermore, the scheme showed that the properties of all "heavy" particles, the so-called *hadrons* (*hadros*: strong) can be explained if they are made of more elementary particles, with the curious property of having an electric charge that is one-third or two-thirds the elementary charge of the electron. Gell-Mann named such particles *quarks*.[2] There are two classes of hadrons: the baryons (*baros*: heavy) such as the proton and the neutron, which are made of three quarks, and the *mesons* (*mesos*: intermediate), such as the pion, made of one quark and one antiquark. A separate category is that of *leptons* (from leptos, light), like the electron and the neutrino, which are unrelated to hadrons and are not made of quarks.

Of course, there were attempts to detect such strange particles with a charge of one-third, but the attempts were unsuccessful. Finally, theoretical physicists found an explanation to why quarks cannot exist in isolation: because quarks must always remain "bounded" to each other. To separate a quark from another quark, it is necessary to invest energy, but the energy must be so great that it would be equivalent to the mass of another pair of quarks. Thus, by injecting enough energy to separate one quark from another, only two additional pairs would be created instead of the original pair (something like cutting a magnet to separate its poles: only two new magnets with their respective poles are produced). The existence of quarks is a clear indication that the concept of mass is not as empirical as it was commonly thought; it can only be interpreted as a parameter in the basic equations... and nothing else. We will return to this issue in the second part of this book.

The quark model explained many properties of the elementary particles, as well as the properties of strong and weak nuclear forces. It is now well established that all hadrons are made up of quarks and antiquarks. To explain how quarks are tied together, it was necessary, once again, to postulate that the interactions between them are mediated by another kind of particle: the *gluon* (not a very original name), which has a role equivalent to that of the photon for electromagnetic interactions.

Initially, the main characteristics of the known particles could be explained with three kinds of quarks: the *up* and *down* quarks, with (+2/3) and (−1/3) the elementary unit of electric charge respectively, and a third *strange* quark with charge (−1/3). Thus, a proton turned out to be made of two *up* and one *down* quarks (total charge 1) and a neutron of one *up* and two *down* quarks (total charge 0). Exotic particles also contain *strange* quarks. More and more exotic particles were discovered in the 1970s and 1980s, and it is now well established that there are six classes of quarks, characterized by what particle physicists call *flavor*: besides the three already mentioned, we have the *charmed, top* and *bottom* quarks (so called for the lack of better names).

A striking property of all six quarks is the huge discrepancy between their masses—or rather, the parameters interpreted as masses in the field equations—. The masses of the *up* and *down* quarks correspond to just one hundredth of the mass

---

[2] Name inspired by a passage from *Finnegans Wake*, by James Joyce: "Three quarks for Muster Mark".

of the proton, so it is believed that almost all the missing mass of this particle is due to the gluons. On the other extreme of the scale, the *bottom* quark is four times more massive than the proton, and the *top* quark has the record with a mass equivalent to 180 protons. So far, there is no explanation of these mass parameters; all that can be said is that they are consistent with the experimental data obtained in large particle accelerators.

As for the theory, the great problem with calculations is that the strength of the nuclear interactions is too strong, and the approximation methods used for electro-dynamics cannot be applied. Nevertheless, it was discovered that these "gluonic" interactions become weak and almost disappear if quarks are densely packed, that is, when the distance between them diminishes, unlike electric or magnetic forces that become stronger at short distances and weaker with increasing distance. One can think of "gluons" as springs that keep quarks bound together but require a lot of energy to be stretched. This circumstance has made it possible to calculate and obtain verifiable results, but with much effort and powerful computers and sophisticated programs. The renormalization techniques are unavoidable since infinite quantities always appear in the calculations.

8.3   The Standard Model

After many years of effort, theoretical and experimental physicists concluded that (almost[3]) all elementary particles and their interactions could be described mathemat-ically by the so-called *Standard Model*, a mathematical model that is based on QFT. It took its final form in the last decades of the twentieth century, bringing together much of what was known about subatomic particles and their interactions. Basically, this model is expressed by means of a Lagrangian function (obviously!), which involves wave functions of particles (quarks, leptons, photons, etc.) and describes the three basic interactions of the atomic world (strong and weak nuclear, and electromag-netic). The formula for that Lagrangian function is "guessed", in a certain way, from the symmetry properties that it must possess. As a first essential condition, it must be Lorentz invariant, that is, it must not change its form when passing from one inertial reference system to another. But, in addition, it must have other properties of symme-tries in abstract spaces in which the equivalent of the coordinates are parameters of the particles, such as spin, isospin, strangeness, etc.

Having this Lagrangian function, the differential equations that describe the inter-actions of the particles are obtained. These are extremely complicated coupled differ-ential equations that can only be approximately solved by very powerful computa-tional programs. So far, it has been possible to calculate some processes that occur between subatomic particles and verify the results experimentally.

In the beginning, the symmetry properties where not enough to include the masses of the particles in the basic Lagrangian. However, theorists found a mathematical

---

[3] A good theory of neutrinos is still missing.

"trick" that consists of including an additional field that would somehow support the masses (but not predict their values). Otherwise, only massless particles could be included in the model. The "trick" worked so well that physicists began to think that this field should be real, and its quantum of energy would be a very massive, neutral and spinless particle (a scalar boson). The famous Higgs field,[4] whose corresponding particle is the Higgs' boson, was experimentally detected in 2012.

The Standard Model has been confirmed in many ways and provides a sufficiently accurate description of nature at the subatomic level…though many physicists are still searching for a deeper theory to support it. It is important to note that the Standard Model is not a theory based on a few axioms. It does not permit to deduce the masses of the particles (not even that of the Higgs particle), nor the mass discrepancy between quarks, and even less the intensity of the interactions between them. All these characteristics must be included in the basic equations as parameters, with their values obtained experimentally. For this reason, the model is sometimes compared— rather unfairly—with that of Ptolemy, which described quite accurately the motion of planets with a superposition of epicycles on epicycles.

In any case, we are very far from the simplicity and "elegance" of the early theories and their ensuing equations that describe the subatomic world.

8.4 Quantum Gravity

When Planck formulated his hypothesis of the quantization of energy, he noticed that the new constant he had introduced (the Planck constant h), together with the other two natural constants, the speed of light c and Newton's constant of gravitation G, could be combined to define natural units of distance, time, and mass. Natural units in the sense that they do not depend on human conventions. The so-called Planck length is given by

$$\sqrt{hG/c^3},$$

and equals about 10^{-35} m, that is, a trillionth of trillion of millimeters, while the Planck time is the time it takes for light to traverse that distance: about 10^{-44} s. As for the Planck mass, it is defined as

$$\sqrt{hc/G}$$

and is equivalent to about 10^{-20} μg. The disparity between these units is notable, since the Planck length turns out to be about 1020 times *smaller* than the typical size of an atomic nucleus, while the Planck mass is about 1022 times *larger* than the mass of an electron. This huge and mysterious disparity has prevented the Planck units from being used in practical measurements.

---

[4] So named for Peter W. Higgs. Nobel Prize 2013 with François Englert.

The Standard Model does not include gravity. To date, despite great and worth-while efforts, it has not been possible to develop a quantum theory of gravitation. A basic difficulty is that the masses of elementary particles are more than 16 orders of magnitude below the Planck mass, while their characteristic sizes exceed the Planck length by about 20 orders of magnitude. Furthermore, the main problem with gravity is that, if one tries to quantize the gravitational field following the same scheme used for electromagnetism, the inevitable infinite quantities appear, but with the aggra-vation that they cannot be eliminated with a renormalization method. Due to this insurmountable problem, several alternative formalisms have been tried, the most famous of them being the "superstring theory".

The basic idea of this theory, at least in its early version, was to describe the most elementary particles as "strings", i.e., one-dimensional objects in place of the usual zero-dimensional points. The vibration of a string should be quantized, and its energy levels would correspond to the various particles. The idea was very attractive, but it did not work as expected and many additional assumptions had to be made to come up with something coherent. In the first version of the theory, it was noticed that the troublesome infinite quantities that always appear could be eliminated if the strings existed in a space of... twenty-six dimensions! Later the number of dimensions was reduced to ten (!) invoking a hypothetical symmetry between bosons and fermions according to a theory (unconfirmed to date) called "supersymmetry".[5]

Where are the extra dimensions? The hypothesis accepted by superstrings prac-titioners was that these dimensions "roll up" on themselves over distances so small that we cannot detect them. It would be analogous to a thread that looks like a one-dimensional line without thickness on a large scale but has a surface with a second dimension "rolled up" that is only noticeable with a magnifying glass.

After many expectations, superstring theory became a new branch of mathematics, extremely complex but perfectly consistent. However, it has never succeeded in proposing any verifiable experiment or explaining any physical process. The main problem is that, to make some quantum gravity experiment, energies of the order of the Planck energy (the Planck mass multiplied by c^2) would be needed, which is a factor of about 10^{16} above the maximum energy that can be achieved with the largest particle accelerator in the world. Superstring theory has revealed the existence of a rich territory in the mathematical world, but without a counterpart in the physical world.[6]

The underlying problem is that it is not even clear that a quantum theory of gravitation would make sense. After all, at the atomic level, gravity is totally irrelevant and could even be a phenomenon that becomes manifest on a large scale only. But, for the present time, one can only speculate.

---

[5] "Supersymmetry" theory postulates a symmetry between bosons and fermions, such that for each fermion there exists a boson and vice versa. For instance, the counterpart of an electron (a fermion) would be an "s-electron" (a boson). However, to date, such supersymmetric particles have not been detected.

[6] For a critical review, see: Peter Woit (2007),Not Even Wrong: The Failure of String Theory and the Search for Unity in Physical Law.

Chapter 9
Cosmology

Se non è vero, è molto ben trovato (Attributed to *Giordano Bruno*.).

Abstract The mathematical description of the evolving Universe was made possible by the general theory of relativity, which predicted the cosmic expansion. The development of the Big Bang Theory is briefly reviewed in this Chapter. The problem of the beginning of the Universe is also mentioned.

When Einstein presented his theory of general relativity, he realized that the concept of a curved space could be applied to the study of the Universe and thus resolve the old dilemma of whether the Universe is finite or infinite. Einstein proposed a relatively simple solution of the equations of his theory that describe a homogeneous universe, finite but bounded, just like the surface of the Earth or any sphere in two dimensions. That would be the case if cosmic space were like the three-dimensional "surface" of a four-dimensional "hypersphere." A spaceship always traveling in a straight line in that hypothetical universe would eventually return to its starting point, just like a cosmic Magellan.

The model proposed by Einstein was a static universe with a homogeneous distribution of matter. But such a universe would collapse by its own gravitational attraction. Therefore, Einstein had no choice but to introduce, without any justification, an ad hoc term in his equations that would physically correspond to a gravitational repulsion at a cosmic scale, thus preventing a universal collapse. This additional term is proportional to the so-called *cosmological constant*, which is usually designated with the letter Λ, and whose value, according to the equations of the theory, would be inversely proportional to the square of the radius R of the Einstein universe: $\Lambda = 1/R^2$.

Einstein's model was static, without change in time. However, the Russian physicist and engineer Alexandr A. Fridman[1] (1888–1925) demonstrated that the same

---

[1] In Russian: Алекса́ндр А Фри́дман.

© The Author(s), under exclusive license to Springer Nature Switzerland AG 2023
S. Hacyan, *The Mathematical Representation of Physical Reality*,
The Frontiers Collection, https://doi.org/10.1007/978-3-031-21254-3_9

equations of general relativity, with or without the cosmological constant, admit more general solutions that represent an expanding universe! Unfortunately, Fridman did not live to learn that the real Universe is in a process of expansion.

9.1   The Expanding Universe

It is only in the last few centuries that astronomers began to get an idea of the real proportions of the Solar System and stellar distances. The first correct measurement of the distance to a star is due to Friedrich Bessel (1784–1846), who, in 1838, deduced that the star 61 Cygni is about 10.3 light years away. Bessel used the triangulation method, based on the diameter of the Earth's orbit and the parallax angle (change in angular position) sustained by the star with a difference of six months.

The triangulation method can be used for a hundred stars within a radius of about a hundred light years, but for more distant stars the parallax angle is too small to be measurable. Much beyond our solar neighborhood, the only method of measuring distance is to compare the apparent luminosity of an object with its intrinsic luminosity. Knowing the two and the ratio between them, the distance is deduced directly from the fact that the apparent luminosity decreases with the square of the distance. Of course, the problem for astronomers is to know the actual luminosity of a cosmic body. Fortunately, stars have various characteristics, such as temperature and chemical composition, that distinguish them from each other, and astronomers have been able to classify them and deduce their intrinsic luminosity according to their category. With this information, it was possible to determine the distances to more and more distant stars.

As for the Universe itself, at the beginning of the twentieth century astronomers were still discussing whether it was restricted to the Milky Way—a very large conglomerate of stars to which our sun belongs—, or if certain luminous spots that were observed with telescopes, the so-called nebulae, were other immense groups of stars, like our Milky Way. The latter hypothesis had been suggested by Kant in one of his early works, noticing that a large distribution of mass would tend to collapse under its own gravitational force, unless it rotates and its centrifugal force compensates the gravitational attraction (as the planets which are kept from falling to the Sun).

The problem of cosmic dimensions began to be resolved with the inauguration, in 1917, of the great telescope on Mount Wilson, in California, the largest in the world at that time and with which hitherto unknown details could be seen. It was with this instrument that Edwin Hubble (1889–1953) was able to detect stars in the Andromeda nebula and could determine that it was two million light-years away. Many other nebulae turned out to be galaxies, conglomerates of billions of stars, even more distant according to Hubble's further measurements.[2] Kant was right!

---

[2] There are also nebulae, like a famous one in Orion, which are clouds of ionized gas within our own galaxy.

Likewise, Hubble noticed that galaxies, like stars, can be classified according to their morphology, with which he was able to determine increasingly larger distances just by identifying the type of each galaxy and deducing its intrinsic luminosity. A new image of the Universe began to appear, with dimensions that exceeded anything previously imagined.

Hubble is also credited with the crucial discovery that the Universe is expanding. He and his colleagues had already noticed that the spectral lines of the light emitted by galaxies were shifted towards the red side of the spectrum. This would imply, according to the Doppler effect, that galaxies were receding from us.[3] Furthermore, when Hubble compared the recession velocity of a galaxy with its distance, he found that the two were directly related and proportional: the further away, the faster! The Universe was expanding, as Fridman had shown mathematically. General relativity thus became a basic and indispensable mathematical tool for the study of the dynamics and evolution of the Universe.

9.2 The Big Bang Theory

If the Universe is expanding, it must have started its expansion sometime in the very remote past. In the late 1920s, the physicist and priest Georges Lemaître (1894–1966) proposed the theory of the "Primordial Atom", according to which the Universe started to expand from a state of a very high density. Nowadays, most cosmologists accept Lemaître's hypothesis, which is based on general relativity and, particularly, the Fridman equations. According to these equations, if the mass density is high enough, the Universe would stop its expansion at some point due to the gravitational pull of the entire cosmic mass and start a contraction. If, on the other hand, the mass density is not sufficient, the expansion would continue indefinitely.

The *Big Bang* theory is the modern version of the Primordial Atom. According to the astronomical observations, the Universe began to expand about 13 thousand million years ago, not only at extremely high densities but also at extremely high temperatures. After that "beginning", the Universe expanded and gradually cooled down. The discovery of the Cosmic Background Radiation (CBR) gave more credibility to the "hot" Big Bang theory, since this radiation was predicted theoretically[4] as what remains of the glow of the cosmic matter when it was at temperatures of thousands of degrees ——the Primordial Fire!——. Since then, the Universe has cooled down to a current temperature of about 3 degrees above absolute zero. This cosmic glow was detected in the form of a microwave radiation by chance in 1965 by radio

---

[3] The Doppler effect shifts the wavelength of light (or sound) emitted by a moving source. The increase of the wavelength is proportional to the speed of the receding source.

[4] The "hot" Big bang theory was advocated mainly by Georges Gamow (1904–1968), Ukrainian physicist who emigrated (fled) to the USA.

astronomers Penzias and Wilson,[5] and since then, it has been possible to measure it with a precision of one part in 100 thousand: it corresponds perfectly to Planck's formula for the blackbody radiation (see Chap.5) at a precise temperature of 2.725 degrees Kelvin, the present-day temperature of the Universe.

As for matter, an unsolved mystery of modern cosmology is the fact that there appears to be much more mass in the Universe than what is observed. For instance, the speed of rotation of the outer stars of a galaxy should decrease with distance, according to Kepler's law, but astronomical observations show that this speed remains constant, which may be due to the presence of a very massive halo of invisible matter around the galaxy. (It has also been proposed that Newton's law might not apply at very large distances... but that is still debatable). Currently, most cosmologists believe that there is a lot of matter in the Universe in the form of elementary particles that are impossible to observe because they do not interact with photons: they interact with ordinary matter only through gravity. To date, there is indirect evidence of the existence of this "dark matter", but it has not yet been detected in a laboratory.

The other great mystery of modern cosmology is the apparent accelerated expansion of the Universe. To date, the finest measurements of the distance to the galaxies is through the observation and use of supernovae as "standard candles." A supernova is a star that suddenly explodes, becoming first as bright as the galaxy that hosts it and gradually fade away over the course of a few weeks. Astronomers have discovered that there is a relationship between the implicit brightness of a supernova and the way it fades away, which is directly observable, thus making it possible to determine the distance to the supernova and its host galaxy. By detecting and studying many supernovae, astronomers were able to refine the cosmological data on how the speed of cosmic expansion varies with distance. For nearby galaxies, the speed is directly proportional to the distance (Hubble's law), but for large distances there must be a slight deviation from that law. The result was a great surprise: the observed speed increased with the distance, when just the opposite was expected. The Universe is undergoing an accelerated expansion.[6]

Within the framework of relativistic cosmology, the only way to describe (but not explain) such an acceleration is by including the cosmological constant – the same originally proposed by Einstein – in Fridman's equations. But the physical origin of this effect remains a deep mystery. It has been suggested that the acceleration force is due to quantum fluctuations of the vacuum, but as we have already seen, the total energy of these fluctuations is formally infinite, and it is not known how to restrict it to something more realistic. For the moment, cosmologists call the cause of the acceleration "dark energy".[7]

---

[5] Arno Penzias and Robert Wilson, American radio astronomers. They received the 1978 Nobel Prize for their discovery.

[6] The 2011 Nobel Prize in Physics was for S. Perlmutter, B. P. Schmidt, and A. G. Ries for "the discovery of the accelerating expansion of the Universe through observations of distant supernovae".

[7] The name of energy is misleading because it is not associated to any mass, but it has been kept that way in the literature.

9.3 Other Theories

There remains the possibility that General Relativity does not apply at very large scales. After all, it has been tested so far for relatively small astronomical scales. Thus, one may wonder whether the Universe is really expanding. Without doubting the data obtained by Hubble, there are physicists who have speculated that the shift of the spectral lines could be due not to the Doppler effect, but to the light losing its energy as it travels through the Universe. However, such a "tired photon" hypothesis has not been confirmed by any experimental data. Another possibility is to accept that the Doppler shift is due to the expansion of the Universe, but according to the Steady State hypothesis, proposed by Hoyle[8] and his colleagues, the density of cosmic matter would remain constant because new matter is continuously being created. There is, however, no evidence nor explanation of such a continuous creation of matter. In any case, the Steady State hypothesis lost credibility after the discovery of the CMB.

From a purely theoretical point of view, all the "non-orthodox" versions of cosmology are consistent to a certain extent, but they have lost credibility in the face of astronomical evidence, and especially due to the remarkable precision of the CMB measurements, in full support of the Big Bang theory. Everything seems to fit perfectly with the theory… provided the two unclear details are taken into account: the existence of invisible "dark matter" and the cosmic acceleration represented by the cosmological constant.

9.4 Time Zero

If the Universe is evolving, there should be a cosmic beginning, a zero of time, but how to define it? According to Fridman's equations, if taken literally, there was a time zero when all the matter was concentrated in a point at infinite density, which is what mathematicians call a *singularity*. In the late 1960s, Roger Penrose and Stephen Hawking proved that the presence of singularities is unavoidable according to the equations of general relativity. Their theorem was based, as all theorems, on a postulate that seemed reasonable: a condition of non-negativity on the energy–momentum tensor that appears in Einstein's equations. However, it was later shown that this condition can be violated within quantum field theory, particularly in the "inflationary" model of the Universe (which we will see below).

In any case, modern cosmologists agree that Einstein's theory of gravitation does not apply to extreme conditions where quantum effects are the dominant ones. A theory that explains gravitational phenomena at the quantum level would be needed to study the origin of the Universe, but, as we mentioned before, a quantum theory of gravity is still lacking. Meanwhile, in the absence of a theoretical unification of gravitation with quantum mechanics, theorists have speculated on the limits of

---

[8] Fred Hoyle (1915–2001), English astrophysicist, also known for his studies of nuclear reactions in stars and some science fiction novels.

validity of Einstein's theory. Clearly quantum effects are not relevant for macroscopic bodies or long enough time intervals, and thus it can be expected that general relativity would still be valid for lengths and times greater than the Planck length and the Planck time. Before the Planck time, which is about 10^{-44} s, phenomena must have occurred that could only be studied with a still unknown quantum theory of gravity, but since this limit is extremely generous, cosmologists have not resisted the temptation to develop theories about the Universe slightly after the Planck time. Shorter times than Planck's are outside the domain of application of modern physics, so it makes no sense to speak of a cosmic "time zero" (except in literature[9]).

Current knowledge about the world of subatomic particles has also made it possible to develop feasible hypotheses about the physical conditions of the Universe in its very early stages. One can resort to some speculative scenarios which attempt to describe extremely high energies, but these energies are still completely outside the range of experimental verification. However, if such speculations are accepted, a consistent picture can be imagined of what the Universe might have been starting from the Planck time... but, of course, only theoretically.

Theoretical cosmologists have resorted to the physical theories developed in the twentieth century, even the most speculative ones, to explain the origin of the Universe. But the basic concept is always that of a field. It is supposed that during the first 10^{-44} s of the Universe, which would be the "Planck era", space and time fluctuated unpredictably, so it would not even make sense to talk about space and time in those conditions. In the beginning, there would be no particles, no light; only the "field". Then, according to a hypothesis proposed by Alan Guth and quite popular among cosmological theorists,[10] the so-called "inflation" occurred, an extremely violent expansion phase of the Universe, with which certain visible properties of the current Universe can be explained. And what produced the "inflation"? a field of course! It was even given a name: the "inflaton".

Does the series of causes and effects end here? One may very well wonder what the cause of the inflaton is, but there is no answer to that. Modern physicists accept the field as the unconditioned, just like the substance of the ancient philosophers.

Summing up, thanks to the great advances in physics and astronomy of the last century, something coherent can be said about the very remote conditions of the Universe and its evolution. Just as the atomic world would be incomprehensible without a mathematical description, the study of the Universe, at the other end of the scale, could not be feasible without the support of a mathematical theory. Modern astronomy and mathematics permit to reach even more remote causes of the world, though a first unconditioned cause will always escape reason.

---

[9] See, e.g., Italo Calvino (1995), *Ti con Zero*.
[10] Alan H. Guth (2000), Inflation and eternal inflation.

Part II

*We [Heisenberg and Bloch] were on a walk and somehow
began to talk about space. I had just read Weyl's book
Space, Time and Matter, and under its influence was proud
to declare that space was simply the field of linear
operations.
"Nonsense," said Heisenberg, "space is blue, and the
birds fly through it."*

        Felix Bloch, *Heisenberg and the early days of
quantum mechanics*. Physics Today, December
1976

Without mathematics, it would be impossible to resolve Zeno's paradox, proposed
to deny the existence of motion or space: "That which is in locomotion must arrive
at the half-way stage before it arrives at the goal."[1] Accordingly, to cover a certain
distance, you must cover first half of it, then half of the rest, and so on ad infinitum.
There is no clear way to refute the paradox with words, but the solution is very simple
with mathematics: To cover a unit of distance, it is necessary to cover, according to
the paradox, a total distance

$$S = \tfrac{1}{2} + \tfrac{1}{4} + \tfrac{1}{8} + \tfrac{1}{16} + \dots,$$

but as one can easily see, $2S = 1 + S$, with the result $S = 1$. The point is that an
infinite sum of terms can be perfectly finite. Zeno cannot fool mathematics!

    The same can be said of the paradox of Achilles and the tortoise. "In a race, the
quickest runner can never overtake the slowest, since the pursuer must first reach the
point when the pursued started, so that the slower must always hold a lead."[2] If v_A
and v_T are the velocities of Achilles and the tortoise, respectively, and L the distance
that the tortoise initially gives as advantage, the distance traveled by Achilles until
he reaches the tortoise is

---

[1] Aristotle, *Physics VI:9, 239b10.*

[2] ibid, *239b15.*

$$L(1 + \nu_T/\nu_A + (\nu_T/\nu_A)^2 + (\nu_T/\nu_A)^3 + \ldots),$$

which is another infinite series but tending to the finite value

$$L\,\nu_A/(\nu_A - \nu_T),$$

just as one would expect from a direct calculation.

Chapter 10
For and Against Mathematics

Abstract The mathematical representation of the world was not easily accepted by all philosophers. Wigner pointed out the mystery of its "unreasonable effectiveness". As suggested by Gödel, mathematics could be an independent reality in the Platonic sense.

About four centuries ago, great thinkers like Galileo, Descartes and Leibniz attempted to create a new and universal science based on purely logical reasoning. Such a program began to take shape with the *Principia*, when Newton proved that the motion of the heavenly and terrestrial bodies can be accurately calculated. This inclusion of mathematics in the foundations of science had, over time, such a success that it would have surprised its very pioneers. Knowledge no longer depended on individual sensations but on mathematical truths, objectively valid for all subjects.

Nevertheless, the new science was not easily accepted in the beginning, not so much because of the intolerance of the time, but because it went against common sense. In fact, it is Aristotle's physics that adjust naturally to our way of apprehending reality. That physics dominated Western thought for many centuries because it was of common sense; it lacked abstractions and provided direct explanations of natural phenomena. It was limited to a description of obvious facts: heavy bodies fall because the center of the Earth is their natural place, fire rises because it pursues the celestial regions, and the planets move on heavenly spheres. It did not have any practical utility but never pretended to.

10.1 Enters Mathematics

Quite on the contrary, Galileo argued in his extensive writings that a power of abstraction is needed to go beyond appearances, and that mathematics may be more appropriate to guide reason than common sense. Newton made full use of mathematics, but he had to rely on concepts of physics that are abstractions and not part of our immediate sensory world. For instance, according to Newton's first law as taught in school, a body always remains in motion, in a straight line and without changing its

S. Hacyan, *The Mathematical Representation of Physical Reality*,
The Frontiers Collection, https://doi.org/10.1007/978-3-031-21254-3_10

93

speed if no force is applied to it. It is the so-called principle of inertia, which seems to be obvious. However, common people know that to move a stone one must push it, and when one stops pushing it stops. Does a stone move indefinitely with just an initial push? To clarify this point, a physics teacher would explain to his students that this principle only applies to frictionless bodies, that is, bodies moving far away from the gravitational influence of planets and stars, but that it does not apply to terrestrial objects that crawl on the ground. But nobody has seen a body moving in a straight and infinite line through cosmic space! It is just an abstraction.

The situation is most evident for the two fundamental theories of modern physics: relativity and quantum mechanics. The former modified our concepts of space and time, and the latter described the atomic world with conceptions alien to everyday experience. Quantum mechanics, more than relativity, does not obey any rational rule and contradicts common sense… but it works perfectly well thanks to its mathematical formulation.

The mathematical language made it possible to go beyond the limits that can be reached with logical or apparently logical reasoning. But there is a deep mystery in it, a mystery that would have astonished Galileo himself. In our time of great scientific discoveries, the efficiency of mathematics to describe the reality of the material world has surprised even scientists.

Eugene Wigner[1] referred to the "unreasonable effectiveness of mathematics"[2] in an article written in 1960 in which he expressed some essential doubts. Why are there mathematical laws that describe Nature? And how do we know that one theory is better than another? There are no obvious explanations. Wigner also pointed out the remarkable fact that mathematical concepts appear in many domains in which they would not be expected. For instance, the famous number π (the ratio between the circumference and the diameter of a circle) appears in most mathematical formulas, such as the Gaussian distribution in statistics, that describe things as foreign to the geometry of the circle as, for instance, the distribution of people by their physical characteristics.

The point is that all the laws of physics apply universally and can be expressed in mathematical terms, at least within certain limits. For instance, massive bodies attract with a force inversely proportional to the square of the distance, which implies that planets and stars move according to that law. But then how do we know the force of gravity does not take any other form in other regions of the universe? We assume so because of the consistency of all the observations and the validity of mathematics, as we know it, in all places in the Universe.

Thus, Wigner concluded: "The miracle of the adequacy of mathematical language for the formulation of the laws of physics is a wonderful gift that we do not understand or deserve", and ended up recognizing: "it is not at all natural that natural laws exist".

---

[1] See Note 27 in Chap. 5.

[2] Wigner (1960).

10.2 Mathematical Platonism

Is mathematics a mere product of the human mind or, on the contrary, does it have an independent existence? The latter is what Plato affirmed according to his disciple Aristotle[3]: mathematics form an intermediate world between matter and the world of Ideas.

We are sure that the Moon exists without depending on our mind or our will. On the other hand, language exists but it is a human product. The Moon may continue to exist without humans, but not language. But what about mathematics? At first sight, common sense would say that mathematics is a product of the human mind, but the fact it describes so well the material world has always intrigued scientists –like Wigner and many more– and has created a sense of mystery about this science. Could there be a mathematical reality, independent of the mind and parallel to physical reality?

In this regard, there are two schools of thought: mathematical Platonism and anti-Platonism. Without arguing in favor of one or the other, it is enough to point out that it will never be possible to reach a conclusion. For instance, Balaguer, in *Platonism and anti-Platonism in mathematics*,[4] shows that the arguments for and against can be perfectly valid. We are in the presence of an antinomy that goes beyond the limits of reason.

For our purposes, let us refer first, briefly, to the anti-Platonist position of philosophers of the so-called neopositivist school, such as Wittgenstein and Carnap.[5] For them, mathematics was a branch of logic that could describe reality, like a language, but lacked its own content. All "mathematical truths" and their proofs could be reduced to a set of symbols that were combined with others according to syntactic rules. One cannot avoid the impression that for these thinkers, mathematics was considered just a method to "calculate" and nothing more.[6] However, for a mathematician, the ability to pose a problem correctly and to find a "trick" to solve it is fundamental; once this is achieved, the calculation becomes automatic (as, I am sure, all mathematicians and theoretical physicists would agree). For Wittgenstein, "the mathematician is an inventor, not a discoverer",[7] but a Platonist would argue just the opposite.

Kurt Gödel (1906–1978), to whom we owe a most important contribution to logic (we will meet again in the next chapters) had an openly Platonist position. In his youth, Gödel frequented the "Vienna circle" and was acquainted with the philosophers of the neo-positivist school, but he found their view too simplistic. In the 1930s, Gödel proved a famous theorem that revealed the limitations of formal languages (to which

---

[3] Aristotle, *Metaphysics*, Chap. IX.

[4] Balaguer (1998).

[5] Rudolf Carnap (1891–1970). Carnap (1928). *The Logical Structure of the World*. See, *e.g.*, §107.

[6] Ludwig Wittgenstein (1889–1951). Wittgenstein (1956). *Remarks on the Foundation of Mathematics*.

[7] Wittgenstein, *op. cit.*, I.168.

we will return in the final chapter). In later years, Gödel though about the philosoph-ical implications of his theorem, but, unfortunately, he rarely decided to publish his work. It is only after his death in 1978 that his manuscripts were rescued, edited, and published, including the drafts of a 1952 lecture on *Mathematical Platonism*, as well as a text criticizing Carnap's thesis on the syntactic nature of mathematics.[8] In those texts, Gödel made a distinction between what he called objective mathematics, which encompasses all true propositions, and subjective mathematics, which includes only those parts that human mind can prove. Unlike the neopositivists, who denied any value to intuition, Gödel claimed that intuition is essential to see a mathematical truth, even if such truth is impossible to prove. Gödel's theorem implies, according to his author, that the human mind performs subjective mathematics, but cannot fully grasp objective mathematics.

Having stated the above, Gödel argued that if mathematics were only a product of the human mind, all mathematical truths could be known and demonstrated, which is not the case at all as shown by his own theorem. A creator is always capable of knowing his own work, and even to modify it, but we cannot fabricate mathematical truths as we please! In conclusion, for Gödel: "[Mathematical] concepts form an objective reality of their own, which we cannot create or change, but only perceive and describe."[9] This statement is very strong, but it is consistent with the Platonic worldview. Let us recall once again (see Chap. 1) that according to Plato, in addition to the material world, there is the world of Ideas or Forms which is not perceived with the senses, but with the mind. It is an eternal and immutable reality in which space does not exist and time does not flow. In the same way, mathematical concepts are also immutable: the Pythagorean theorem, for instance, is valid eternally and anywhere in the Universe. Accordingly, Gödel argued that mathematics is as objective as the material world: "… [M]athematics describes a non-sensual reality, which exists independently both of the acts and [of] the dispositions of the human mind and is only perceived, and probably perceived very incompletely, by the human mind."[10]

Moreover, there are assertions in mathematics that seem clear enough but have not yet been proved or have been proved with great difficulties. Examples of the latter are the "four color map" theorem—four colors are enough to illuminate any map—and the "last Fermat" theorem—the equation $a^n + b^n = c^n$ has no solutions with integer numbers, except for $n = 1$ or 2—. Each one has been proved with considerable effort and after many years of struggling, even recurring to computer for the four color theorem. The "Goldbach conjecture", which can be expressed in very simple terms—every even integer number (except 2) can be expressed as the sum of two prime numbers—has been checked for billions of numbers, but until today, it has not been proved. One can wonder whether these theorems "exist" in the Platonic World, where they are simply "true", like the Moon which exists but is not easy to reach.

---

[8] Feferman et al. (1995); *Some observations about the relationship between theory of relativity and Kantian philosophy*. In: *Kurt Gödel: Collected Works. Vol III*. S. p. 202.

[9] Feferman *op. cit.*, p. 320 (30).

[10] Feferman *Op. cit.*, p. 323 (38).

Gödel recognized that his Platonic realism was not very popular among his colleagues. Years later, the editors of his manuscripts admitted that his point of view had become rather more popular after his time "... in some measure because of his advocacy of it, but perhaps more importantly because every other leading view seems to suffer from serious mathematical or philosophical defects."[11] In summary, mathematics exists, according to Gödel, in the same way that objects or ideas exist. With our minds, we can explore and describe them within the limitations of our mathematical language. For instance, we can deduce (or guess?) the form that the Lagrangian function should have to describe the interactions in the atomic world, which is not just a calculation (as Wittgenstein believed). Mathematics is part, perhaps the most accessible part, of that other non-sensuous reality: the world of Ideas. It is for this reason that Plato required to know geometry to enter his Academy.

---

[11] *Idem,* pp. 303–304.

Chapter 11
Against Mathematics

Ceci n'est pas une pipe.
Magritte

Abstract Despite its successes, Newtonian physics was criticized for its excess of mathematical abstraction. For instance, Goethe developed his own theory of colors to refute Newton's optics. The main criticism against the objectivity of physics is related to its lack of subjectivity.

A century after its appearance, Newtonian mechanics had been established as a paradigm to be followed by science. Thus, Kant affirmed that "there is only as much genuine science as there is mathematics".[1] Of course, it is a matter of definition, but what Kant had in mind was a science of Nature that should have "a pure part that is the foundation for the empirical part and is based upon a priori knowledge of natural things." The condition for such a priori knowledge, according to Kant, is that it can be expressed in the mathematical language. Chemistry, for instance, was for Kant "a systematic art", although that is no longer the case for modern chemistry, which is based on experimental facts and supported by the theory of quantum mechanics.

11.1 Against Newton

Despite the undeniable success of deciphering Nature with mathematics, it is understandable that there were reactions against the attempt to rigidify science. Since the beginning of the nineteenth century, many distinguished thinkers rejected the reductionism of the world to formulas and equations. A famous engraving by William Blake depicts Isaac Newton sitting naked in the middle of a sea-like landscape,

---

[1] In the preface of *Metaphysical Principles of the Science of Nature*.

© The Author(s), under exclusive license to Springer Nature Switzerland AG 2023
S. Hacyan, *The Mathematical Representation of Physical Reality*,
The Frontiers Collection, https://doi.org/10.1007/978-3-031-21254-3_11

absorbed in tracing geometric figures on a sheet of paper, oblivious of his surroundings. The most accepted interpretation of the image is that it symbolizes the aridity of the abstract world of Newtonian science compared to the richness and variety of the real world.

A notable example of the rejection of mathematical descriptions is Goethe's, who strongly criticized Newton's theory of light and his famous experiments reported in his *Optiks*. The great German poet was also interested in the phenomena of light, but his approach was very different. He tried to reproduce Newton's experimental results but failing to do so, he concluded that the English scientist had been a charlatan and he decided to develop his own theory of colors.

Goethe spent several years performing his experiments in optics and published the whole of his results in 1810, in a vast two-volume treatise entitled *Farbenlehre*, with more than 900 printed pages. Regardless of the validity of his theory, Goethe did not hide his distrust for scientific abstractions, which may be useful on describing Nature, but no more, and he concluded: "It does not occur to the architect to pass off his palaces as mountain sides and forests".[2]

Goethe's theory of colors was analyzed by Werner Heisenberg[3] with the purpose of understanding his rejection of traditional optics. The co-founder of quantum mechanics accepted that Newton's theory, based on mathematical proofs, is foreign to what we perceive directly with our senses. The real world, the one we observe without intermediary instruments, has an extensive variety and color that, for most people (and even more so for artists), cannot be reduced to a dull mathematical treatise in which colors are identified with numbers. There is no doubt that the Newtonian theory describes the phenomena of light correctly, but it is of little use to a painter who seeks to understand and represent reality. Actually, the problem of human color perception is quite complex and Newtonian optics is not sufficient to explain the physiological mechanism that permits seeing the world in color. In this sense, Goethe's theory could be closer to reality, if color is to be understood as a subjective perception.[4]

According to Heisenberg, Newton's and Goethe's theories deal with two different levels of reality: an objective and a subjective one.[5] The latter is the reality of Goethe's theory.

---

[2] Quoted by Heisenberg (1990): *Goethe's view of Nature and Science,* in *Across the Frontiers.* Chap X.

[3] Heisenberg, *op cit.*

[4] For a more current review of Goethe's theory, Ribe and Steinle (2002).

[5] Heisenberg, *op. cit.*

For his part, Schopenhauer, who deeply admired Goethe and supported his theory of color, was also a stern critic of the use of mathematics. For him,

> ... calculating is not understanding and in itself does not afford a comprehension of things... It can even be said that *where calculating begins, understanding ends*; for whoever is occupied with numbers is, while calculating, a complete stranger to the causal connexion and to geometrical construction of the physical sequence of events.[6]

We will return to Schopenhauer in the following chapters.

11.2 Subject and Object

Criticisms of the use of mathematics, such as those just mentioned, are mainly due to its limited applicability to the study of subjects. In general, a theory that claims to be scientific must be objective and universal. Objectivity means that the object of study is independent of what an investigator thinks; for instance, the motion of the planets, described by Newtonian physics, does not depend on the personal views of the astronomers, nor can they influence their course. As for universality, this refers to the fact that a scientific theory must be valid in all cases without exceptions. Thus, Newton's theory of gravitation applies to all massive bodies, but it would be of no use if it did not apply to all celestial bodies without exception. In other words, the laws of Nature must be, in principle, the same here and now, in any part of the Universe and at any time... or at least that is what we must assume if we want something like science to exist! We will return to the matter in the following Chapters.

Likewise, for scientific research to make sense, we must be sure of a fundamental premise: Nature does not deceive. Planets rotate in elliptical orbits, and we can be sure that they will never move otherwise on their own "will". In any case, a scientist may come across a phenomenon that does not fit known theories, but it will never be the case that Nature is playing a trick; instead, it will be concluded that the theory is not correct. In brief, the goal of a scientific theory is to reconstruct a picture of the Universe that is coherent and objective. Thanks to mathematics, science has become objective because it was possible to quantify the world and describe it with numbers that are the same for all human beings, unlike subjective factors that cannot be described with mathematical formulas.

What happens if the object of study is the human mind? How do we know that the subject is not "lying"? Lying may even be unconscious, due to "false" memories or "wrong" interpretations. Apparently, this was one of the great concerns of Sigmund Freud, who insisted that psychoanalysis, his personal work, should be a science. However, he could not appeal to objectivity, because there is no such thing as an "objective thought", independent of the self-conscious beings. While it is possible, in principle, to relate some mental states to physicochemical processes in the brain, that does not explain why a person thinks so or so. There is an irreducible element

---

[6] The citation is from Schopenhauer, 4R, Chap. IV, § 21 (Sch.'s emphasis). See chapter 14 below.

of free will, which undermines any claim to predict individual human behavior in a quantitative way.

The abandonment of subjectivity led some thinkers to reject an excessively scientific view. Edmund Husserl, in *The Crisis of the European Sciences*, saw the moment in which science abandoned subjectivity in the seventeenth century, precisely when Galileo discovered the power of mathematics to describe and understand Nature..[7] Despite his unquestionable successes, the new science, according to Husserl, "excludes those questions most vehement to the human being: questions about the meaning (or lack of meaning) of his entire human existence." In this sense, Aristotle's physics would be more human because it is consistent with everyday experience. A universal science should encompass both the subjective and the objective; otherwise, Husserl claimed, we will always be restricted to a single (two-dimensional) plane, in which we can move in all directions; so another dimension is needed to rebuild the human universe.

Likewise, Paul Feyerabend, in *Farewell to Reason*[8] and in his posthumous work *Conquest of Abundance*,[9] criticized the enthusiasm and expectations aroused by scientific progress. Like Goethe, and later Husserl, Feyerabend thought that the excessive abstraction of modern science had caused the loss of a more direct contact with the world. In brief, he believed that modern science is only one of the possible visions of the world, and not necessarily the best one. Without denying the great technological advances, Feyerabend insisted that the spiritual problems of individuals cannot be solved by scientific reason. It can be argued that his radical position is due to an artificial confrontation between two worldviews that, in fact, can complement each other. In this respect, it is significant that Feyerabend, like Husserl, attempted to vindicate the philosophy of Aristotle, who, unlike his teacher Plato, avoided abstractions and preferred a direct view of the world. For this reason, an entire chapter of *Farewell to Reason* is dedicated to Aristotle's physics, so discredited today. Another chapter is dedicated to the Austrian physicist and philosopher Ernst Mach, a central figure of positivism that was in vogue at the end of the nineteenth century, who advocated a physics based on concrete experiments and not on mathematical theories. Feyerabend also tried to vindicate this physicist and philosopher of science who was quite popular in his time but was forgotten by the next generations.

Thus, we have the situation that the mathematical program has been extremely successful and exceeded all expectations in the physical domain. But, on the other hand, due to the difficulty to deal with subjectivity, the "scientific" mathematical approach has not produced any far-reaching result in those disciplines related to the mind.

Having clarified all the above, let us return to what we could describe as "objective sciences", in which (apparently!) the subject does not intervene, and mathematics plays an important role.

---

[7] Husserl (1954).

[8] Feyerabend (1975).

[9] Feyerabend, (1999).

Chapter 12
Programs of the Mind

Abstract The debate between rationalism and empiricism is briefly reviewed: Is the mind born in blank or with a "software" (in modern terms)? Kant distinguished between a priori and a posteriori knowledge and claimed that space and time are a priori forms of perception of our cognitive apparatus. It is pointed out here that his thesis is consistent with modern physics.

Learning is remembering. In *Meno*, the Platonic dialogue, Socrates questions whether virtue is learned or innate. In favor of the latter notion, he argues that there are ideas coming from the very deep "memories" of the mind. The existence of such memories, or innate ideas, is most evident in the case of mathematics, Socrates says. To prove his point, he asks his host to call a young slave to whom he wants to propose a problem of geometry. The young man had no previous education, but he was able to solve the problem with a minimal assistance of Socrates, who intervened with only brief comments to guide him. The conclusion drawn by the Athenian philosopher was that the psyche of man is immortal, and that the slave had only remembered what he already knew without being aware of it.

12.1 Rationalism and Empiricism

Psyche is usually translated as the soul, although, in the context in which Plato uses it, it is rather the mind. Should this passage be interpreted as Plato's argument for the transmigration of the souls? Or was the disciple of Socrates referring to hereditary mental structures? Could these be something like the instinct with which a bird builds a nest or a spider spins its web? Would it be, using the modern language of computing, a *software*, that is, a pre-established "program" that we are born with? Apart from these speculative interpretations, the fact is that man is the only animal capable of having abstract ideas, and among these the mathematical ideas stand out, as Socrates remarked.

Centuries later, Descartes, in his *Meditations on First Philosophy*, also reflected on how our mind apprehends an external reality, independent of the subject. He concluded that the idea of God is innate, because otherwise we could not explain how we conceive an Infinite Being from our finite experience. And if we accept this fact, we can also accept that other ideas are equally innate. These innate ideas are what allow us to understand the world, and mathematical ideas occupy a special place among them, since.

> Physics, Astronomy, Medicine and all other sciences which have as their end the considera-
> tion of composite things, are very dubious and uncertain; but that Arithmetic, Geometry and
> other sciences of that kind which only treat of things that are very simple and very general,
> without taking great trouble to ascertain whether they are actually existent or not, contain
> some measure of certainty and an element of the indubitable. For whether I am awake or
> asleep, two and three together always form five, and the square can never have more than
> four sides...[1]:

Finally, the creator of the analytical geometry came to the conclusion that "bodies are not properly speaking known by the senses or by the faculty of imagination, but by the understanding only".[2] A position that is identified with *rationalism*,[3] in contrast to *empiricism* where ideas are built from experience only.

The English philosopher John Locke had a contrary opinion. In his famous *Essay on Human Understanding*, written at the end of the seventeenth century, he insistently argued that there are no innate ideas, but that all ideas arise from two sources: the experience of the world, captured with the senses, and the reflection of these ideas that we carry out with the mind. Thus, the human mind is born as a *tabula rasa,* something like a clean slate on which ideas are marked from experience and reflection. Accordingly, Locke criticized Descartes' thesis that the idea of God is innate, since not everyone conceives a Supreme Being. On the contrary, Locke pointed out that primitive peoples worship a plethora of deities with human forms and passions, and less primitive but poorly educated people imagine God as a man sitting in the heavens.[4] The idea of God, Locke argued, arises only from the observation and the study of His creation.

Locke claimed that there are simple ideas acquired through sensation and reflection, and complex ideas that the mind builds from the simple ones. In particular, he doubted that natural philosophy could become a science, since our ideas of material objects are due to what we perceive with our limited senses and, at most, some ancillary apparatuses.[5] Thus, without denying the importance of Newton's work, he did not see in it more than a coherent structure built with the ingenious union of axioms and propositions of geometry.[6]

Leibniz read Locke's treatise with great interest and, although praising its author for his insight, he wrote his own "new essays" on the same subject, *Nouveaux essais sur l'entendement humain*, where he commented and criticized, line by line, the ideas

---

[1] Descartes, *Meditations on First Philosophy; First Meditation.* Translated by E. Haldane.

[2] Ibid. Second *Meditation.* Ibid.

[3] Something like the "archetypes" and the "collective unconscious" of Jung, mentioned in Chap. 1.

[4] Locke, *Essay...* Ibid, Book 1, Chap. III (15–17).

[5] Locke, Ibid, Book 4, Chap. XII (10).

[6] Ibid, Book 4, Chap. VII (3).

of the English philosopher. Leibniz did not believe that the human being is born with the mind as a *tabula rasa*. As a proof of innate ideas, he sustained Descartes' argument that the idea of God was impressed on our minds. Furthermore, he cited arithmetic and geometry as examples of innate ideas, as Plato had done in the dialogue we mentioned above. Leibniz also disagreed to natural philosophy having restricted limits; perhaps physics would not be a perfect science, he argued, but we must not give up the effort to perfect it. Just as geometers have not proved all the axioms, but continue to deduce important results, so too "physicists, by means of some principles of experience, will give reasons for countless phenomena and even predict them in practice".[7]

David Hume (1711–1776) is often identified with the empiricists, although he explained, in *An Enquiry Concerning Human Nature,* that the discussion about the existence of innate ideas was due to philosophers who did not have clearly specified what they meant by idea. Unlike Locke, he emphasized the difference between pure ideas and direct perceptions. For instance, the perception of burning is not the same as the memory of a burn, nor is the idea of love the same as being in love. Thus, Hume distinguished two classes of perceptions of the mind that differ in their intensity: *ideas*, which are usually vague, and *impressions*, so called for the lack of a better name, which would be living perceptions.[8] Our ideas would be copies of our impressions that are clear and lively.[9] It is in this sense that Hume concluded: "But admitting these terms, *impressions* and *ideas*… and understanding by *innate*, what is original or copied from no preceding perception, then may we assert that all our impressions are innate, and our ideas not innate".[10] (This, by the way, allowed Hume to identify the instinct of animals with impressions: a subject matter that neither Locke nor Leibniz considered.)

As for human reason, Hume distinguished two ways of operating: one is by relating ideas to each other and the other is by recognizing matters of fact. An example of the former is mathematics, where the relationship between various objects of reason can be demonstrated rigorously and without doubt by means of theorems (for instance, relating the squares of the legs to the square of the hypotenuse in a right triangle). On the other hand, the problem with matters of fact is that we cannot be sure that they will occur or not; unlike the theorems of geometry, they cannot be rigorously proved, and their truth is not self-evident. Thus, Hume raised a fundamental question: how can we be sure that this or that fact happens? For him, the answer laid in the cause-effect relationship: we know, by custom, that a certain cause produces a certain effect, and that effect is due to that cause. Also, some cases can be safer than others. For instance, we have no doubt that the Sun will rise tomorrow because it has always done so, but we are only reasonably certain that it will rain tomorrow. As we will see later, the issue of the cause-effect relationship, already clearly addressed by Leibniz, will be fundamental in the philosophy of science.

---

[7] Leibniz, *Nouveaux essais*… Book IV. Chap. XII, § 9 (my translation).

[8] Hume, *An Enquiry*… Sect. II § 12.

[9] Hume, *ibid.*, Sect. VII, § 49.

[10] Hume, *op cit.* Sect. II §17.

Hume shared with Locke the skepticism about the scope of our knowledge of Nature, even with the help of mathematics.[11]

> Every part of mixed mathematics proceeds upon the supposition that certain laws are established by nature in her operations; and abstract reasonings are employed, either to assist experience in the discovery of these laws, or to determine their influence in particular instances, where it depends upon any precise degree of distance and quantity [...] Geometry assists us in the application of this law [of motion], by giving us the just dimensions of all the parts and figures which can enter into any species of machine; but still the discovery of the law itself is owing merely to experience...

Summing up, human reason for empiricist philosophers could not go very far, even with the help of mathematics, since it is inevitably based on sensory experience. Newton's work, for instance, was just a curiosity, very ingenious, but of little use for "understanding"—whatever that means—the world.

12.2 Kant

Thus, in brief, rationalist philosophers believed that there are innate ideas known a priori, that is, before any experience, while for empiricists, all ideas are acquired a posteriori, after experience. Of course, much is learned from experience, but the question remains whether the mind is born as a *tabula rasa*. In this regard, Kant's *Critique of Pure Reason* (CPR)[12] is a fundamental text that begins with the following statement "There can be no doubt that all our knowledge begins with experience", but Kant immediately clarifies (B 1):

> But though all our knowledge begins with experience, it does not follow that it all arises out of experience. For it may well be that even our empirical knowledge is made up of what we receive through impressions and of what our own faculty of knowledge... supplies from itself.

Thus, Kant set out to investigate in the CPR the limits that reason can reach in its attempt to understand and interpret the world perceived with our senses. But it must first be recognized that such a perception, according to Kant, is based in our concepts of space and time, which are given a priori. That is, space and time are prior to experience: they belong to the *innate* mental structures with which we interpret what is perceived with the senses.

In the first chapter of the CPR, entitled *Transcendental Aesthetics*, Kant asserted that space and time are *forms of perception* (*Anschauungsformen*: from *Anschauung*: view, *Anschauen*, to look at). Objects affect our senses as phenomena in space and time, and phenomena are produced by a *thing-in-itself*, which is not directly accessible to the senses. This is where Kant proposes his revolutionary thesis: space and

---

[11] Hume, *op cit*. Sect. IV §27.

[12] All subsequent references to Kant's CPR follow the academic numbering. I follow the English translation by Norman Kemp Smith (Macmillan, 1929).

time are alien to the thing-in-itself, but we perceive phenomena through these forms of perception.

Next, Kant set out to establish the basic principles with which we can process and understand those perceptions received from the outside world. In the following chapters of the CPR, collected under the title *Transcendental Logic,* he wrote (A50, B74):

> Our knowledge springs from two fundamental sources of the mind; the first is the capacity of receiving representations (*Vorstellungen,* receptivity for impressions), the second is the power of knowing an object through these representations... Intuition (*Anschauung*) and concepts (*Begriffe*) constitute, therefore, the elements of all our knowledge...

Thus, Kant distinguished "the science of the rules of sensibilities in general, that is, aesthetics, from the science of the rules of the understanding in general, that is, logic." (A52 B76).

In other words, we receive information from the world through our senses and the forms of perception, space and time, but that information has to be processed by the mind to make sense of it. For instance, I can hear a foreign language, but if I am not acquainted with it, I only hear meaningless sounds. Another example: a blind man who regains his sight in adulthood will only see luminous spots, since his brain has not learned to process that information, as it normally happens from the very early childhood. In the modern language of computation, it is as a computer receiving information as a succession of zeros and ones—corresponding to the passing or blocking of electrons in its circuits—, but a program is needed to order and produce something meaningful from a sequence of zeros and ones.

Kant asserted that our understanding works with pure concepts, given a priori, which are integrated in our mind. He grouped them into four forms of understanding– quantity, quality, relation and modality– with three subdivisions of each. Without going into detail, let us give some examples:

Judgment of universal quantity. "All cats are mammals".

Negative quality judgement. "Cats are not herbivores".

Hypothetical relationship judgment "If it meows, so it must be a cat".

Problematic modality judgement. "That cat could be Bob".

Etcaetera (see the CPR for the full table: A70 B95, but be warned that Kant did not give any example).

Next, continuing with the same structure, Kant proposed to classify "the pure concepts of synthesis that the understanding contains within itself a priori" (A80 B106), the very general concepts which Aristotle called categories, and with what we join a particular concept (subject) with another concept (predicate) to form a judgment (synthesis). In parallel with the forms of understanding, Kant classified four categories (quantity, quality, relation, modality) with their respective three subdivisions. Thus, for instance, the category *relation* is divided into:

(1) substance and accident (the cat is meowing)

(2) causality and dependence (the cat is meowing because it is hungry),
(3) reciprocity of action and reaction (the cat catches the mouse, the mouse is caught
      by the cat).

12.3   Kant and Modern Physics

Can the above principles be sustained in the light of modern physics, where space
and time are fundamental and unavoidable concepts? In an earlier text,[13] I argued
that Kant's conception is perfectly consistent with our current understanding of the
quantum world: the concepts of space and time, as they are known in the macroscopic
world, do not apply to the atomic world due to the existence of non-local effects
and the non-existence of a past-future temporal order. This fundamental and well-
established difference from the macroscopic world reinforces Kant's conception of
space and time as forms of perception with which we perceive phenomena. As we
know today, all phenomena emerge, in some way, from the level of the atomic world.

Theoretical physics describes the world with mathematics, mainly with differential equations in which time appears as a simple parameter that can run both in
one direction or another. The direction of time is manifested only in systems with a
huge number of atoms, that is, objects of the macroscopic world. As we mentioned
before, reversing the direction of time is equivalent, according to quantum mechanics,
to interchanging particles for antiparticles and reflecting the process in a mirror (the
CPT theorem, see Chap. 7). The only law of macroscopic physics where a distinction
between past and future appears is the second law of thermodynamics, according to
which the physical quantity known as entropy never decreases. However, it is a statistical law, in the sense that it does not apply to single atoms but to a large number of
particles; it only implies that time is vastly more likely to flow from past to future
than vice versa. For instance, if water spills from a glass on the floor, it is highly
unlikely—but not impossible!—that each water molecule, just by chance, evaporates
back into the glass simultaneously.[14]

Furthermore, the subjective time we perceive directly does not have a direct relationship with the time that appears in the equations of physics. As Eddington well
noted from the beginnings of relativity:

> If it is found that physical time has properties which would ordinarily be regarded as contrary
> to common sense, no surprise need be felt; this highly technical construct of physics is not
> to be confused with the time of common sense.[15]

As for the space of common sense, it is not the space that appears in the equations
of quantum mechanics. In the quantum world, the observation of one particle can
affect instantly the properties of another particle at an arbitrarily large distance. As we
explained in the chapter on quantum paradoxes, such an interaction (which Einstein

---

[13] Hacyan (2004) and (2006).

[14] For a more detailed discussion, see: Hacyan (2004), Chap. 4.

[15] Eddington (1923) *The mathematical theory of relativity*, Chap. 1.

once called "spooky") can occur regardless of the distance between particles, as if the communication travelled at infinite speed, thus violating a fundamental principle of relativity. But it must be remembered that relativity is a theory of the macroscopic world and so such faster-than-light interactions only appear at the atomic level.[16]

---

[16] Hacyan (2004) and (2006), *op. cit.*

Chapter 13
Geometry a Priori

> [M]an has an instinctive tendency, not rooted merely in external
> experience, to interpret his sensory perceptions in terms of
> Euclidean geometry.
> Wolfgang Pauli (1949)

Abstract The mathematical discovery of non-Euclidean geometries apparently
refuted Kant's doctrine of space and time. However, Euclidian geometry could well
be part of our cognitive apparatus. Poincaré claimed that geometry cannot be the
object of experiments. The implications for general relativity are discussed here.

If space is a form of perception, geometric ideas should also be given a priori. Consequently, according to Kant, the axioms of geometry would be synthetic judgments a
priori. Synthetic in the sense that they join two ideas or concepts that are independent
in principle, but that reason relates to each other. For instance, we have an a priori
idea of a triangle and an angle; their synthetic union couples these two concepts in a
theorem of geometry: the internal angles of a triangle always add up to 180 degrees.
We reach this conclusion from a mental process, and not because we experimentally
measure the angles of all the triangles that can be drawn or constructed.

13.1 Geometry and Physics

In Kant's time, the only known geometry was that of Euclid. Although Kant never
mentioned the great Greek mathematician, he was obviously not aware of other
geometries, such as those proposed by Bolyai and Lobachevsky which appeared
after his time. Starting from this obvious fact, some authors were rash to conclude
that the existence of non-Euclidean geometries disproved that space is given a priori,
since the true geometry of space—be it Euclidean or curved—would be the object
of experience. Lobachevsky once suggested that astronomers could check the type

of geometry by measuring the angles of a triangle formed by two positions of the Earth and some star, and see if it they really add up to 180 degrees.[1]

The argument that non-Euclidean geometries contradict Kant's idea of space seems to go back to Helmholtz.[2] In an essay written in 1870, Helmholtz argued that the axioms of geometry cannot be synthetic judgments a priori, since they can be subjected to experience. As Lobachevsky had suggested, it would be possible to determine by means of measurements—terrestrial or astronomical—the sum of the three angles of a triangle or see whether two parallel rays of light always maintain the same distance between them. Thus, the axioms of geometry would be empirical and not a priori products of the mind.

Helmholtz developed his own version of a non-Euclidean geometry based on what he believed must be the fundamental condition of all geometry: "the possibility that figures move without changing their shape or size"; that is, it must be possible to translate and superimpose one figure (a triangle, a circle, etc.) on another to check whether they are equal or not. Otherwise, it would be impossible to define what a measure is if a ruler were to change its size by moving from one place to another. According to Helmholtz: "the axioms of geometry are not concerned with space-relations only but also at the same time with the mechanical deportment of solidest bodies in motion". In any case, Helmholtz was aware that a strict Kantian could argue that the stiffness of bodies is a property that must be assumed a priori in any measurement. "But then we should have to maintain that the axioms of geometry are not synthetic propositions, as Kant held them: they would merely define what qualities and deportment a body must have to be recognized as rigid".

In this regard, Poincaré questioned how the equality of two figures can be deduced. Obviously, moving one until it coincides with the other, but how to move it without deforming it? This does not solve the problem, said Poincaré, since "it would not make any sense for a being that inhabits a world in which there are only fluids".[3] The possibility of such a motion for an ideal solid body (not a natural one) cannot be proved experimentally and must be accepted as a postulate.

In any case, the theory of relativity did away with the concept of a rigid body. A rigid motion cannot be defined within the framework of Einstein's theory, due to the simple fact that it is not possible to move "simultaneously" all parts of a solid body. At the very most, a push at one end would produce a compression wave that would propagate through the body at the speed of sound in the medium. Furthermore, what is defined as the length of a body depends on the reference frame in which it is observed, and the measurement of that length has an absolute value only in the system in which the body is at rest.[4] In summary, it makes no sense to invoke the material rigidity of bodies as the basis for a geometry that intends to describe the

---

[1] Jammer, 1993, p. 149.

[2] Helmholtz (1995), "*On the origin and significance of geometrical axioms,* in *Science and Culture,* p.244.

[3] Poincaré (1952), *Les géométries non euclidiennes,* in *La science et l'hypothèse.*

[4] In fact, the motion of an extended (not a point-like) body remains a problem not entirely solved in the theory of relativity, both special and general.

physical world. It is only in the mathematical realm that the rigid displacement of a figure can be ideally defined. In fact, mathematicians use the concept of congruence as the idealized translation of a geometric figure to compare it with another figure in another position.[5]

Arguments like those of Helmholtz were taken up by Reichenbach in his critique of Kant's concepts of space and time.[6] Basically, his argument was that the nature of space must be verified empirically from the motion of material bodies in a gravitational field, and therefore cannot be taken a priori as Euclidean.

But Kant never claimed that *physical* space was Euclidean. Rather, he argued that space is a pure form of perception, that is, an a priori component of mental structures that permits us order our experiences and make sense of them. In this regard, Kurt Gödel wrote a long essay on Kant and his relations with the theory of relativity. Gödel argued that the theory of relativity and, in general, modern physics are consistent with the ideality of space and time. In particular, he supported the argument that space is a form of perception in accordance with Euclidean geometry:

> In the case of geometry, e.g., the fact that the physical bodies surrounding us move by the laws of a non-Euclidian geometry does not exclude in the least that we should have a Euclidian "form of perception", i.e., that we should possess an *a priori* representation of Euclidian space and be able to form images of outer objects only by projecting our sensations on their representation of space, so that, even if we were born in some strongly non-Euclidean world, we would nevertheless invariably imagine space to be Euclidean...[7]

Accordingly, it can be said that Euclidean geometry is an a priori part of our form of perception, and, in that sense, it is the natural geometry with which we perceive the world. An obvious proof is that it is impossible to visualize a curved space of three or more dimensions: it can only be conceived through its mathematical representation. As for a curved two-dimensional space, such as the surface of a sphere, we can visualize it but only as embedded in the usual three-dimensional space. In other words, three-dimensional Euclidean geometry is the innate representation with which we perceive space, although mathematics permits the conception of other geometries without limits.

The important point that thinkers like Reichenbach missed is that Riemannian geometry is necessarily based on Euclidean geometry, since it is its generalization.[8] This can be seen from the very definition of Riemannian space, which presupposes, *as a fundamental condition*, that the space can be approximated as a flat space in a sufficiently small region. More precisely, a Riemannian space must, by definition, admit a locally Euclidean tangent space, and also a definition of the distance between points. The surface of the Earth, for instance, is curved but it appears to be flat

---

[5] Helmholtz's formulation of geometry is a primitive version of what was later developed in mathematics as the theory of Lie group (see Hacyan, 2009).

[6] Reichenbach (1958).

[7] Gödel, *Some observations about the relationship between theory of relativity and Kantian philosophy.* In: Feferman (1995); p. 241.

[8] A classic technical book on the subject: Eisenhart, (1959), *Riemannian geometry.*

(Euclidian!) in a small enough region, which makes it possible to elaborate city maps without drastically altering the distances.

In summary, Euclidean geometry is the natural geometry adapted to our form of perception. The possibility of conceiving more general spaces does not refute Kant's argument that space is a form of intuition given a priori; on the contrary, it proves the immense capacity of mathematics to transcend our conceptual structures.

13.2   Geometry as an Object of Experience? Poincaré

Let us return to the question of whether the axioms of geometry can be verified experimentally. In an essay written in 1898, two decades before Einstein's general relativity, Poincaré argued that geometry is a convention, since there is no point in performing something like a "geometric experiment":

> Let it be a material circle, let its radius and circumference be measured, and see if the ratio of the two lengths equals π. What have we done? We have done an experiment, not on the properties of space, but on those of the matter with which this *roundness* has been achieved, and those of the ruler used for the measurements.[9]

Something similar would happen with astronomical observations. For instance, regarding Lobachevsky's suggestion to determine the curvature of space, Poincaré noted:

> What we call a straight line in astronomy is simply the path of a ray of light. If, then, we were to discover negative parallaxes, or to prove that that the parallaxes were greater than a certain limit, we would have the choice between two conclusions: we could give up Euclidean geometry or modify the laws of optics and admit that light does not rigorously propagate. in a straight line.

He then concluded: "It is not necessary to add that everyone would regard this [latter] solution as more advantageous".

It must be remembered that Poincaré wrote this essay when the usefulness of non-Euclidean geometries in physics was still unknown. He was wrong on this last point: fifteen years later, Einstein showed that Riemann's geometry is more advantageous for the description of gravity.

Apparently, Poincaré's point of view was refuted by the theory of general relativity and its successful confirmation. Thus, his position fell into oblivion due to the then widely accepted belief that the Riemannian nature of space-time makes physical sense and, as such, is open to experimentation. Unfortunately, the untimely death of Poincaré did not allow him to know the general theory that Einstein presented in its

---

[9] *L'Expérience et la Geométrie,* in Poincaré (1968). My translation.

final form in 1916.[10] Surely, he would have had a very fruitful discussion with its author.

Overall, the question remains why the Euclidian geometry seems to be the natural one? Poincaré found the answer in natural selection, thanks to which "our spirit adapted itself to the conditions from the outside world, it adopted the most advantageous geometry for the species; in other words, the most comfortable". There is no doubt that this geometry is the most comfortable, since it was accepted as the natural one for two millennia, but Poincaré never explained why this was so. A Kantian, however, would argue that it is more convenient because it is given a priori. In any case, Poincaré's conclusion is clear: "the geometry is not certain, it is advantageous". Certainly, Euclidean geometry is the most advantageous to use, except to describe gravity in the relativistic regime, for which Riemannian geometry is more advantageous. But is Riemannian geometry indispensable for this task?

Had Poincaré known Einstein's new theory, he would have possibly contended that this theory simply implies that light rays do not propagate along "straight lines" due to the gravitational attraction of some massive body. Empirically, it is equivalent to supposing that light moves along "null geodesics" in curved space, as postulated in general relativity, or along a path curved by gravity in a Euclidean space: there is no way to distinguish between the two interpretations, and therefore general relativity does not contradict the conventionalism of geometry.

Nevertheless, a crucial question is whether a relativistic theory of gravitation requires a curved space. As already mentioned, the basic equations of this theory were also obtained by Hilbert, but he had to rely on the concept of a Riemannian space proposed by Einstein earlier. Now, if neither Einstein nor Hilbert had thought of that basic idea, it would have appeared most probably with the formalism of *gauge theory* formulated in the 1950s.[11]

Basically, gauge theory permits to deduce (one could say: "guess") the equations that describe an interaction from its basic symmetry properties. In other words, symmetry is enough to determine the dynamical laws! (Recall, as seen in Chap. 8, that by symmetry we mean that physical properties are not changed by displacements or rotations, either in real or in abstract spaces.) The idea was first used by C. N. Yang and R. L. Mills in 1954 to obtain the equations that describe the nuclear interactions, considering the symmetry that we mentioned earlier, viz. that purely nuclear interactions do not change if protons and neutrons are exchanged with each other. This formalism was fully generalized a year later by Ryoyu Utiyama (1916–1990) in an article[12] in which he showed that the equations obtained from gauge

---

[10] A preliminary versión was Einstein's 1915 article: "Die Feldgleichungen der Gravitation". Sitzung der physikalische-mathematischen Klasse, *25*, 844-847. The final version of the theory was published in 1916 as "Die Grundlage der allgemeinen Relativitätstheorie". Annalen der Physik. IV Folge. 49.

[11] For an extensive review of gauges theories of gravitation, see Blagojevic and Hehl (2013).

[12] Utiyama (1956).

theory for the Lorentz transformations have exactly the same structure as the equations of a Riemannian space of four dimensions. In particular, a tensor with the same mathematical structure as the Riemann tensor appears.

Utiyama's result proved that there is an exact correspondence between Riemann's four-dimensional geometry and the gauge theory associated with the Lorentz transformations of special relativity in the Minkowski space. To get the complete general relativity with this formalism, the presence of matter must be included in some way. This can be achieved directly from Hilbert's principle of least action,[13] thus yielding the Einstein equations of general relativity without assuming the curvature of space-time but interpreting the metric tensor as a potential of the gravitational field.

In summary, Utiyama's work on gauge theory, combined with Hilbert's principle of least action, showed that Riemannian geometry is highly advantageous for the theory that describes this fundamental interaction. Einstein's great merit was to advance a theory of gravitation through a very ingenious analogy with Riemannian geometry. But, as Poincaré could say, the use of this geometry is a very convenient convention, but a convention anyway.

In fact, every prediction of general relativity can also be interpreted as taking place in a Euclidean space, since Einstein's equation can be interpreted as the equations for a field described by a metric tensor[14] in flat space. Mercury's perihelion shift can be ascribed to the corresponding corrections of Newton's law of gravity, and the bending of light can be interpreted as the interaction of light with a gravitational field. Even the formation of black holes can be explained in classical mechanics, simply noticing that light cannot escape from the gravitational attraction of a sufficiently compact and massive object... as Laplace already knew. Nevertheless, it is undeniable that the interpretation of the gravitational field as a curvature of space-time is an attractive and convenient description of this fundamental force of nature.

As for the metric tensor, it can also be interpreted as a mathematical function that describes a special field in a Euclidean space. In fact, that is the direction followed implicitly (although they do not state it explicitly) by the first researchers of general relativity, such as Eddington and Tolman, in their books on general relativity (see Chap. 4).

As for cosmology, the expansion of the Universe is a prediction of the dynamical Fridman equations which, in turn, are derived from general relativity. Similar predictions can also be deduced from other, more complicated theories of gravitation, although the general relativistic version is the clearest one. Furthermore, it is now a fully accepted fact that astronomical observations are consistent with the Universe being spatially flat on average; that is, a space which is "stretching" over time, but not curving (like a flat rubber sheet being stretched). A restriction, therefore, to a "flat but expanding" geometry could be sufficient.

Clearly Einstein was convinced of the non-Euclidean nature of space, but his conviction was based on the fact that, during his lifetime, the only known way to

---

[13] See Chap. 4.

[14] That is, a spin 2 field.

unify gravity and relativity was with the mathematical tool of Riemannian geometry. However, as it can be seen in an article Einstein wrote in 1949 (for a book dedicated to him and edited by A. P. Schilpp[15]), he was well aware that the problem was considerably more difficult than what Helmholtz or Reichenbach had thought. He even imagined a dialogue between Poincaré and Reichenbach on the subject "Is a geometry—looked at from the physical point of view—verifiable (viz. falsifiable) or not?", with both scientists giving their arguments, but without reaching a clear conclusion for the reader.

---

[15] Schillpp (1949). See also Hacyan (2009).

Chapter 14
Causality, Sufficient Reason and Determinism

Dios mueve al jugador y éste la pieza.
¿Qué dios detrás de Dios la trama empieza...?
Ajedrez
J. L. Borges *(God moves the player, and he, the piece/ Which god behind God begins the plot...? (Chess))*

Abstract As pointed out by Leibniz and Schopenhauer, the principle of sufficient reason is the condition for the understanding of Nature. It is related to causality, by which we associate a stimulus to its perception. Determinism and uncertainty are also considered in this Chapter.

The principle of sufficient reason is an unavoidable topic of discussion in philosophy. Leibniz stated this principle in the following way[1]:

> By virtue of the principle of sufficient reason, we consider that no fact would be true or existing, no true statement, without there being a sufficient reason for it to be so and not otherwise.

Schopenhauer also delved into the subject in his famous doctoral thesis, *The fourfold root of the principle of sufficient reason* (4R)[2] and argued that this same principle is inextricably associated with causality—the cause-effect relationship—of which four roots can be identified, two of them related to the perception of matter. The first root refers to the causal relation between a physical phenomenon and our way of perceiving and interpreting it. The second root has to do with logical processes: the relationship in which a concept or idea is deduced from another that is its cause. The third root refers to mathematics, particularly geometry, and the process in which geometrical truths are deduced from postulates as their causes. Finally, the fourth root refers to the cause for which a certain act is undertaken, that is, the motivation (conscious or not) to act in a certain way.

---

[1] Leibniz, *Monadologie.* (My translation).

[2] *Ueber die vierfache Wurzel des Satzes vom zureichenden Grunde* (1813).

© The Author(s), under exclusive license to Springer Nature Switzerland AG 2023
S. Hacyan, *The Mathematical Representation of Physical Reality*,
The Frontiers Collection, https://doi.org/10.1007/978-3-031-21254-3_14

14.1 Causality

Sometime later, in his main work, *The World as Will and Representation* (WWR), Schopenhauer argued that causality is another way of perceiving phenomena, different from space and time, but is also given a priori, that is, before any experience. Our mind is so structured that we always seek for the cause of every effect. Schopenhauer thus opposed the empiricists, such as Locke and Hume, who argued that our concept of causality occurs a posteriori, after the accumulation of all the empirical experiences according to which we see that every effect has a cause.

For empiricists, causality must be a probabilistic concept, since we cannot know from experience that a phenomenon will cease to occur in the future. For instance, the Sun rises every morning, but will it rise tomorrow? At the best, we know from experience that it is immensely more likely that the Earth will keep rotating than to collide with an asteroid.

According to Schopenhauer, causality is related a priori to the principle of sufficient reason since we believe that any effect must have a cause which is its reason. In his doctoral thesis (4R), he identified the first root of the principle of sufficient reason as the cause-effect relation between the stimuli by the senses and the phenomena that produce them; it is an essential relationship that permits to process all the information received through our senses, mainly sight and hearing. He gave as an example the physicochemical stimuli produced by light on the retina and the transmission of this information to the brain through the optic nerve. This raw information must be further processed by the brain to produce an image of the world. In the modern language of computing, we could say that the brain has a software that processes the received information and produces images of the world, just like a computer which transforms a myriad of small electric currents in its circuits into images on a screen (without the correct program, we only see a succession of zeros and ones). According to Schopenhauer—and anachronism apart—, our mind is equipped with an innate software with which we learn to use through experience.

To put the matter in its historical context, it must be remembered that in Schopenhauer's time, the perception and study of Nature was still done, in many cases, directly through the senses, as he discussed extensively in 4R. Besides telescopes and microscopes, it is mainly in the nineteenth century that scientists began to study the world with the extensive support of sophisticated devices designed to peer into natural processes. But Schopenhauer's argument about the way information is processed appears to be even more valid: indeed, the relation between the observer and the observed object is given through the intermediacy of all sorts of devices, and the data must be interpreted according to some well-established program. That is, there is a causal relation between a physical process and the effect that follows from processing the data. Gone are the days of the direct perception of nature, so celebrated by poets!

In this same respect, Pierre Duhem (1861–1916) noted a century ago:

An experiment in Physics is not simply the observation of a phenomenon; it is also the theoretical interpretation of this phenomenon.[3]

As Duhem rightly pointed out, a layman visiting a physicist's laboratory would not guess what is being done there, since he would only see a set of devices and monitors displaying graphs and numbers. In his time, Faraday performed all his crucial experiments with apparatuses that made visible the effects produced by the invisible electric and magnetic forces, but, above all, he was able to interpret his results and draw conclusions from his observations. Likewise, modern physicists who study subatomic particles do not see atoms or electrons (much less quarks, Higgs bosons, etc.!) but only tracks on photographic plates and numbers; they can only analyze the data collected by their detectors and deduce their cause according to a theoretical model.

Thus, there are several steps in the processes of understanding a physical process. According to Kant, there is a thing-in-itself that produces the phenomena, which are perceived as stimuli in our sense organs or in more complex apparatuses. In the former case, as Schopenhauer pointed out, the relationship between cause and effect is provided by the innate mental structures (given a priori) in the brain; in the latter case, the relationship is given by artificial structures, such as the software of an artificial brain, which processes the data according to a pre-established theoretical model. In modern science, the experimental results are necessarily interpreted according to a theory that provides the causal relationships between the raw data and the theoretical description. Without a theory, the results of an experiment in physics would be only an incomprehensible compilation of numbers and graphs. As Einstein said, "it is theory which decides what we can observe" (see Footnote 5 in Chap. 5).

14.2 Sufficient Reason

A simple example of the use of sufficient reason is given by the concept of force, to which we return one more time. According to Newton's second law, a body changes its state of rest or uniform motion if a force acts on it. But this is the only way to define force: as the sufficient reason for a body to change its state. Euler proposed in his treatise on mechanics a simple definition of what force is *not*:

> Clearly, there is no reason why [a body] should move to this or that place. Consequently, in the absence of a sufficient reason to keep it moving, it must remain permanently at rest.[4]

The sufficient reason for motion is what we call force.

Let us see another example of the principle of sufficient reason, more complex and closer to our times. The phenomenon of superconductivity was discovered in

---

[3] P. Duhem (1914), *La théorie physique.*

[4] ...[P]erspicuum est nullam esse rationem, quare potius in hanc vel illam plaguem moveatur. Consequenter ob defectum sufficientis rationis, cur moveatur, perpetuo quiescere deben. (Euler, *op cit.*, corollary to proposition 7).

1911 by Kamerlingh Onnes,[5] who made the unexpected discovery that the electrical resistance of mercury falls exactly to zero below a certain critical temperature: an electric current flows freely without any resistance. Physicists spent several decades studying this phenomenon until they found its explanation, that is, its sufficient reason, in quantum mechanics.

Yet another example is the current situation in cosmology: all the observational data fit quite well with the model of the Universe predicted by the theory of general relativity, but only if it is assumed that there is a mysterious invisible "dark matter", together with a cosmic acceleration due to the so-called "dark energy", which is even more mysterious. Today, most cosmologists believe that the model is correct and look for direct evidence of the dark matter as a sufficient cause of what is observed. As mentioned in Chap.4, other researchers have pointed out that gravitation may not behave, on very large scales, as predicted by general relativity. In any case, a sufficient reason is sought for all observations, either in the form of a new kind of matter or a correction with respect to general relativity.

In the examples above, causality is closely related to the principle of sufficient reason: there must be a reason for every phenomenon. If an unexpected result appeared, it may be caused by something yet unknown within a known theory (such as dark matter) or a limitation of that theory (such as a possible modification of general relativity). The purpose of science is to find out what is the cause of an unforeseen phenomenon. No scientist would be content with a lack of explanation for a newly discovered phenomenon.

14.3 Determinism

Causality is found in all domains of physics and, as Schopenhauer pointed out, it is a direct consequence of the principle of sufficient reason, which the subject applies to the objective world. The principle of causality:

> consists in the fact that, if a new state of one or more objects appears, another state must have preceded it, from which the new state follows regularly, that is, as long as the first state exists... the first state is called cause, the second effect.

From which he concluded: "the law of causality refers solely and exclusively to changes in material states, and nothing else".[6]

Causality, however, must not be confused with determinism. Determinism in physics refers to the fact that an effect can be calculated from a known cause. More precisely, the state of a system at time t_1 determines unambiguously its state at later time t_2. In this sense, it can be said that a state, at a certain time, is the cause of another state at a later time.

However, the above definitions of causality and determinism apply with various subtleties in physics, classical and modern. Indeed, both Newtonian and relativistic

---

[5] Heike Kamerlingh Onnes (1853–1926). Nobel Prize 1913.

[6] 4R, §20.

mechanics are causal and deterministic in principle, but they may not be deterministic because their evolution depends inevitably on the initial conditions of a system.

To that respect, going back to the early nineteenth century, it is worth mentioning another important (and more terrestrial) treatise by Laplace, *Theory of Probabilities*, which addressed the problem of handling a large set of data that cannot be known with absolute precision.[7] Due to this limitation, Laplace held that we can only resort to statistical notions to establish the probability that a certain phenomenon will occur. Time proved him right since indeterminism was found at the heart of Newtonian mechanics itself.

Almost a century after Laplace, Henri Poincaré set out to study an apparently simple problem, to which Laplace had found only an approximate solution. It is about the motion of three massive bodies that attract each other according to Newton's simple law of gravitation. Poincaré found that the problem hid a mathematical structure of a previously unsuspected complexity. His study gave rise to what decades later, with the help of computers, became known popularly as the "theory of chaos".

The underlying problem is that not all physical systems can be analyzed with precision, no matter how corrects the basic theory is and how powerful the computational methods are. Many systems, such as liquids, gases, or societies of living beings, are very sensitive to initial conditions, to the point that a very small change of the cause produces a huge change of the effect, and the whole process becomes unpredictable. Some examples are the turbulence in water and the behavior of the climate. It is now evident that the overall behavior of a very complex system cannot be reduced to a few laws that govern its parts. In other words, the whole is greater than the sum of its parts. Even if we understand how an air molecule interacts with other similar molecules, that knowledge is useless for predicting something as common as the formation of clouds.

14.4 Uncertainty

Uncertainty is not the opposite of determinism. Determinism means that a given cause has always a unique effect… which can or cannot be calculated in practice. Uncertainty deals with the fact that the cause or the effect may not be known with absolute precision.

In quantum mechanics, the evolution of an atomic system is determined by the Schrödinger equation, which permits to calculate the evolution of the wave function that describes all possible states of the system. However, according to the standard interpretation, the observation forces a quantum system to appear in one of those states in a non-deterministic way.

---

[7] Laplace is often credited with the worldview that everything is determined by the laws of physics. This is, however, an out-of-context interpretation of a passage in his treatise on probabilities. See Hacyan (2004), chap. XI.

As for uncertainty, it is related in quantum mechanics with the famous Heisenberg principle that refers to the impossibility of measuring both the position and the momentum of a particle (in the same direction) with absolute precision.

As explained in Chap. 5, many processes in the atomic world can be observed and measured, at least in principle, with limitless precision. It is just a matter of choosing two observables whose observations do not affect each other. Thus, the real implication of the uncertainty principle is that not all measurements can be meaningful and not all questions to Nature can be answered without contradictions. However, this does not contradict the principle of sufficient reason, since this principle does not imply that a cause or an effect must be determined with absolute certainty.

In this regard, in an essay on philosophical issues, Heisenberg (1969)[8] tried to refute Kant by arguing that the principle of sufficient reason does not apply to quantum mechanics. He gave as an example the decay of a radioactive nucleus: the exact time in which the nucleus emits an electron cannot be calculated; therefore, the process is strictly probabilistic. However, Heisenberg was confusing determinism with sufficient reason. Indeed, it is not possible to predict the exact moment at which a nuclear decay will occur, but that is not the point. The point is that physicists could search the reason for such a process… and, in fact, they found it. It was then realized that the reason for a nucleus to emit an electron is the beta decay: the spontaneous transmutation of a neutron into a proton, an electron, and an (anti)neutrino. Later, physicists discovered that the reason for this process is the existence of a weak interaction that takes place in the nucleus. And still later, the reason for such an interaction was explained by the now well-established theory of electroweak interactions, based on the existence of the massive W and Z bosons. And the reason for the existence of such particles and their interactions may, perhaps, be found in the future with a more fundamental theory than the current Standard Model. And so on…

14.5   The Unconditioned

If every effect does have a cause, does it make sense to look for the most initial cause, a cause which is not the effect of a previous cause? Kant showed that, regarding this question, there were two possible lines of thought for pure reason, which he related to the thesis and antithesis of his antinomies in CPR. The former consists in stopping at some point and postulating that we have found the first and "unconditioned" cause, which we accept without demanding any further explanation; the latter consists in always looking for a cause that precedes another, without ever reaching a basic and irreducible one.

Kant pointed out that our reason feels more comfortable in the first path, that is, accepting an original and irreducible cause that requires no further questioning. On the other hand, the second path—always looking for a previous cause—is perfectly

---

[8] Heisenberg (1972), *Quantum Mechanics and Kantian Philosophy*, in *Physics and Beyond*.

valid as a logical method, but it does not satisfy reason, which gets tired of the endless search for deeper and deeper causes.

In the perspective of modern physics, it seems that reason is content to accept that quarks and electrons—and perhaps superstrings—are the most fundamental constituents of matter, and also that the first cause of the Universe is a field which cannot be reduced to something more fundamental, and that we cannot say anything about what happened before the Planck time. However, the second path is also valid and cannot be ruled out. Thus, the Big Bang could be a particular event in a larger "universe" , in which a particular event caused the formation of "our" Universe. All this, due to a field, the "inflaton" , whose existence may one day be explained by something more basic… but, of course, only through mathematics in the mathematical world.

*

The underlying argument of philosophers like Kant and Schopenhauer is that we study and describe the world with a priori concepts. Thus, for instance, ancient philosophers were interested in the vague concept of substance as the substratum of the world, and modern physicists developed the mathematically precise concept of the "field", interpreted as the ultimate underlying substance of the world. The field is mathematically described in terms of space and time variables and is interpreted as the sufficient reason for the existence of subatomic particles, that is, matter and its interactions.

As for the Universe, does it make sense to look for something like its ultimate cause? In terms of cause-effect relations: every phenomenon has a cause, which in turn is produced by another cause, and so on. The question, then, is whether there is a chain of causes and effects that ends in a final cause that does not depend on any prior one. But nothing prevents us from asking what the cause of that ultimate cause was, and so on…

Chapter 15
Matter and Causality

Abstract Kant considered space and time as forms of perception, but he did not include matter. Schopenhauer argued that matter is perceived through causality and the principle of sufficient reason. Accordingly, he elaborated a table of "predicables" which includes space and time together with matter, in substitution to Kant's transcendental logic.

Schopenhauer devoted an entire section of WWR to Kant's philosophy. He acknowledged that "Kant's greatest merit… [is] the distinction of the phenomenon from the thing-in-itself",[1] and went as far as affirming the uniqueness of the thing-in-itself, which he interpreted as "the Will" (but we will not dwell into this matter as it falls outside the scope of the present text).

For our purposes, let us note that Schopenhauer fully accepted Kant's doctrine on the ideality of space and time, and agreed with him that "transcendental aesthetics cannot contain more than these two elements [space and time]"(B58).

In contrast, Schopenhauer believed that Kant's "Transcendental Logic" exposed in the CPR, was unfunded. He attributed his four classes of categories (as we saw in Chap. 12) to the great philosopher's taste for "architectural" structures. For Schopenhauer, all the categories could be reduced to only one: "I demand that we throw away eleven of the categories, and retain only that of causality".[2]

Furthermore, Schopenhauer included matter as another form of perception, different from space and time, which manifests itself through causality and the principle of sufficient reason. Let us see in the present chapter to what extent this principle, as well as matter, can be considered a priori concepts in the light of current knowledge.

---

[1] Schopenhauer, WWR Appendix.

[2] WWR, Appendix.

S. Hacyan, *The Mathematical Representation of Physical Reality*,
The Frontiers Collection, https://doi.org/10.1007/978-3-031-21254-3_15

15.1   Matter

If space and time are forms of perception, what about mass, which is the other fundamental concept that appears in physics? Everything physical is measured in terms
of the basic units of space, time and mass—such as meter, second and kilogram—.
However, Kant did not consider mass as a pure form of perception. On this point,
he was quite explicit in stating that "the possibility of the synthesis of the predicate
'weight' with the concept of 'body' … rests on experience" (B12). In other words,
matter is quantified as mass, and mass can be determined empirically, either directly
as weight on a balance, or indirectly by the motion of a body (according to Newton's
second law).

Kant did not dwell long into the matter. He pointed out that although the conservation of mass is empirically well established, it is nonetheless a postulate.[3] He
illustrated this point with the example of a philosopher who wants to determine the
weight of smoke: "Subtract the weight of the remaining ashes from the weight of
the burned wood, and you have the weight of the smoke". However, this conclusion is based on the postulate that "matter (substance) does not disappear, but only
undergoes an alteration of form" (B 228).

Thus, the perception of mass is related to the principle of its permanence understood as a postulate. Even if a process were found by which mass were not
conserved and this principle were disproved experimentally, some sufficient reason
for the discrepancy would be sought, without abandoning the general principles. For
instance, mass is transformed into energy, according to modern physics.

As an example of the principle of permanence, it was discovered in the 1930s that
certain radioactive nuclei emitted electrons. When the energy of these particles was
measured, it turned out that a certain amount was missing. Niels Bohr then proposed
that the principle of conservation of energy might not apply, but Pauli conjectured
that the missing energy was carried away by an invisible and unknown particle. Pauli
was right: a few years later the neutrino was discovered. Thus, Pauli followed the
reasoning of Kant's hypothetical philosopher mentioned above: instead of smoke,
he weighed a new particle. But what would have happened if Bohr had been right?
In that case, it would have been necessary to search the reason for such a violation
of a physical law… which would have caused a revolution in physics.

Let us return to Schopenhauer. He pointed out that causality—the cause-effect
relationship—necessarily manifests itself in matter, or more precisely, in the change
of state of material objects. Causality makes no sense in an empty space or without
the passing of time. For this reason, Schopenhauer thought that matter is, in a certain
sense, a form of perception such as space and time since: "only as *filled* are these
[space and time] *perceivable*. Their *perceivability* is *matter*…".[4] And furthermore:
"Matter [is] the perceptibility of time and space, on the one hand, and causality that
has become objective, on the other"(4R, § 35).

---

[3] For a more extensive discussion, see Hacyan (2004) Chap. 9.
[4] 4R §18, original underlining.

Later on, Schopenhauer detailed this point. For instance, Chap. 4 of WWR is devoted to the concept of matter:

Time and space, however, each by itself, can be represented in intuition even without matter; but matter cannot be so represented without time and space.

And in the same chapter, he specified that "causality relates solely and entirely to the determination as to what kind of state or condition must appear *at this time and in this place*" (original underlining).

In summary, Schopenhauer's conclusion was that matter is what we perceive directly with our senses, not space or time. We perceive space through the matter that fills it and the passage of time with a material clock (or our own body) that occupies a place in space.

The foregoing can be made clearer if we consider the table that Schopenhauer elaborated and named *predicabilia* (predicables[5]) in Chap. 4, Vol. 2, of WWR, "On Knowledge a Priori".

15.2 *Predicabilia* a Priori

For the reasons previously mentioned, Schopenhauer included matter along with space and time in his table of *predicabilia* a priori, which he defined as "all fundamental truths rooted in our a priori knowledge of perception"(WWR Ch. 4, Vol 2). It is as if the mind were made up of boxes or files, forming a broader structure, and in which each idea, in the form of subject and predicate, could be accommodated. What Schopenhauer pointed out is the parallelism and similarity between the predicables that correspond to space, time, and matter, which he presented in three parallel columns. In his own words:

Matter… is not *object* but *condition* of experience, just as are space and time. This is why, in the accompanying table of our pure fundamental knowledge *a priori*, *matter* has been able to take the place of *causality*, and, together with space and time, figures as the third thing which is purely formal, and therefore inherent in our intellect.

Some of the predicables in his table may be anachronic, but a few are worth examining and briefly commenting to understand what Schopenhauer had in mind. They illustrate the similarities and also the differences, as forms of perception, between space and time, on the one hand, and matter on the other hand.

Thus, for instance, the first predicable is: "There is only one time…—only one space…—only *one* Matter, and all different materials are different states of matter; as such it is called *Substance*". Of course, the concept of substance used by ancient philosophers is now obsolete, but what Schopenhauer had apparently in mind is a substratum of all matter. We will return to this subject a few lines below.

---

[5] *Predicable*: one of the five most general kinds of attribution in traditional logic that include genus, species, difference, property, and accidents. (Merriam-Webster)

The second predicable: "Different times are not simultaneous but successive—Different spaces are not successive but simultaneous—Different matters (materials) are not so through substance but through accidents". Here, he returns to the concept of matter as an accident of what he calls substance, something that we do not perceive directly but through the phenomena.

The third a priori predicable says; "Time cannot be thought away...—Space cannot be thought away...—The annihilation of matter cannot be conceived, yet the annihilation of all its forms and qualities can". Indeed, space and time are the unavoidable parameters in the equations of physics that describe all physical processes: space and time appear as variables, while mass always appears as a *constant* parameter in these equations. Even if abstract spaces can be conceived in mathematics, physicists will settle in the end for a description in terms of space, time, and the parameter mass. As for the annihilation of matter, the modern interpretation is that it is equivalent to the concept of energy, and the total mass-energy cannot be annihilated (for instance, the annihilation of an electron with a positron produces two gamma rays, particles of pure energy).

The fourth a priori predicable is: "Matter exists, i.e., acts in all the dimensions of space and throughout the whole length of time, and thus unites and thereby fills these two. In this consists the true nature of matter. It is therefore through and through causality."Here, Schopenhauer insists on what he had already expressed about matter.

The fifth predicable a priori is that space, time and matter are infinitely divisible. As regards space and time, if they are forms of perception and not physical entities, it makes no sense to ask whether they are divisible or not, just as it makes no sense to ask whether an idea is divisible. The situation is different for matter. The existence of quarks, electrons and other elementary particles is currently well established, but they are described as point particles in physical theories, with no structure and therefore no further division. Although the hypothetical "superstrings" have some structure, it is in an abstract mathematical space. If a deeper structure of quarks and electrons were to be discovered in the future, scientists would not stop at that level but continue to search for even more elementary constituents. But it would be an endless search.

The sixth predicable is that space, time and matter are homogeneous and form a continuum. This does not apply to elementary particles as such, but the field of modern physics, as a substratum of matter, does form a continuum (see next section). There have been some mathematical attempts to quantize space or time, but without any clear result.

Finally, let us discuss in some detail the eighteenth predicable:

Time is not measurable directly through itself, but only indirectly through motion, which is in space and time simultaneously; thus time is measured by the motion of the sun and of the clock.

Space is measurable directly through itself, and indirectly through motion, which is in time and space simultaneously; thus, for example, an hour's walk, and the distance of the fixed stars expressed as so many light years.

Matter as such (mass) is measurable, i.e., determinable according to its quantity, only indirectly, thus only through the *magnitude of the motion*, which it receives and imparts by being repelled or attracted.

All measurements involve space, time, and mass. In the twentieth century, a closer relationship between space and time was revealed by the theory of relativity, together with the equivalence between mass and energy. A further connection between mass and time was revealed by Planck's postulate of the quantization of energy, thus establishing a new constant of nature. But the irreducible elements of all measurements are the same three: space, time, and mass.

Time was measured in the past by the periodic motion of the heavenly bodies, and it is measured today by the periodic vibrations of a cesium atom; there is a perfect and consistent relation between that time and all the objects in the physical world (or, at least, such a relation must be assumed). As for space, it was initially measured with respect to the human body (foot, inch…), then with respect to the Earth when the metric system was instituted in the French Revolution, and finally today in terms of the distance traveled by light in a (already defined) unit of time. For mass, however, it has been considerably more difficult to find a standard in terms of natural constants. It was only recently that this goal was achieved fixing the value of the Planck constant by definition (see Chap. 5).

In any case, mass can only be measured through motion (or equilibrium) in space and time, as Schopenhauer stated in his table of predicabilia. Which reminds of the old discussion, started by Euler, about whether the primary concept was mass, force, or acceleration.

Hence, from mass to force, and from force to the new concept of energy, we return to mass at the beginning of the twentieth century, when Planck discovered a relationship between energy and frequency, and Einstein realized the equivalence between mass and energy. The circle of successive definitions was expanded to three members: force, energy, and mass, which are entirely equivalent to space, time and mass.

15.3 Substance and Matter

Let us take a closer look at the concept of substance with which we have come across repeatedly. The classical philosophers called "substance" the underlying and permanent element of the world. Kant thought it necessary to presuppose its existence since, otherwise, "experience would never be possible if we were willing to allow that new things, that is, new *substances*, could come into existence" (B 229).

As for Schopenhauer, he agreed that the different manifestations of matter are accidents of the substance. On this point, at least, he follows Locke, who divided qualities into primary and secondary, the latter being mutable. However, Schopenhauer argued in his critique of Kantian philosophy (appendix to WWR) that "…the concept of *substance* was formed merely in order to be the vehicle for surreptitiously introducing the concept of the immaterial substance", namely the soul. Substance is simply reduced to matter if the concept of soul is eliminated.

Of course, physics does not deal with the soul, and therefore it is irrelevant, from our present point of view whether substance is identified with matter. In any case, the conservation of mass-energy and the field are the fundamental concepts in modern physics since they encompass both matter and its (immaterial) interactions. It should be clear that the human mind needs to rely on the concept of something more general than common matter to justify its existence, which could be the substance or, presently, the field.

On the other hand, according to the second predicable cited above, material bodies do not manifest themselves through substance, but through their accidents, which would be the phenomena they produce. So, in a modern language, "accidents" can be interpreted as particular manifestations of the fields that physicists deal with.

Is there an ultimate substance of things? As far as is known in modern physics, it would be the quantized field: the vibrations of the field are the stuff that particles are made of. The quantized field, unlike the substance of the ancient philosophers, is divisible and quantifiable. Could substance be a primitive antecessor of the idea of field? The correct question could rather be whether concepts such as substance or field correspond to some mental image that is part of our cognitive apparatus, and to which modern physics has found a realization through mathematics?

Chapter 16
Mathematics and Reality

Il y a longtemps que personne ne songe plus à devancer l'expérience, ou à construire le monde de toutes pièces sur quelques hypothèses hâtives. De toutes ces constructions où l'on se complaisait encore naïvement il y a un siècle, il ne reste plus aujourd'hui que des ruines.

Henri Poincaré, 1897 (*L'Analyse et la Physique*, in *La Valeur de la Science*, Poincaré (1970). It has been a long time since no one thinks of getting ahead of experience, or of constructing the world from scratches based on a few hasty hypotheses. Of all these constructions with which people still naively delighted a century ago, only ruins remain today. (My translation))

Abstract Can the laws of physics be deduced from pure mathematical reasoning? Eddington's attempt to determine the values of the fundamental constant is an extreme example. The meaning and possibility of a "Theory of Everything" is examined in this Chapter.

Schopenhauer was not fond of mathematics. Without denying its practical use, he was convinced that mathematics, though it can produce a quantitative description of the material world, could never provide an understanding of its causal relationships. "Where calculating begins, understanding ends" was his statement on this subject (4R, 21; see also WWR Chap. XIII). His point of view may seem out of scope, but as we saw in the first part of the present book (particularly Chap. 10), his aversion to mathematics was shared by other intellectuals of his time who yearned for a direct perception of nature, without the intermediacy of abstract concepts.

Nowadays, we are used to the enormous success of mathematics in describing physical phenomena. However, Schopenhauer was right in a sense; a mathematical description only produces numbers, but not an understanding of the phenomenon that is being observed. Quantum mechanics is an excellent example of this: it is based entirely on abstract concepts —wave functions, operators, probability amplitudes, spin, etc.—, together with a rigorous mathematical formalism. But although quantum mechanics has proven to be the most accurate description of physical phenomena, any attempt to "explain" it in terms of familiar concepts inevitably leads to paradoxical conclusions.

S. Hacyan, *The Mathematical Representation of Physical Reality*,
The Frontiers Collection, https://doi.org/10.1007/978-3-031-21254-3_16

The fact that mathematics is so effective may well be one of the greatest mysteries of modern physics. Such effectiveness is, in fact, quite unreasonable (as Eugene Wigner rightly pointed out). In any case, we can agree with Schopenhauer that "Where calculating begins, understanding ends", provided we also realize that calculating can reach unsuspected limits!

16.1 Theory of Everything

Although the effectiveness of mathematics is well established, its limits are difficult to establish. As there are philosophers like Schopenhauer who distrusted its scopes, at the opposite extreme there are physicists and mathematicians who believed that something like a "theory of everything" could eventually be achieved, namely a theory that would explain the fundamental properties of the most elementary particles by means of sophisticated mathematical concepts, and with a minimum of empirical data.

In any case, it is necessary to clarify what is meant by a "theory of everything", since there is no unanimity among physicists about the matter. For most of them, it would be a mathematical theory that explains, from a few mathematical postulates, the fundamental interactions of nature and some (or perhaps all!) natural constants (electron charge, quark masses, spin, etc.). It would be a wide generalization of Maxwell's theory of electromagnetic interactions, extended to electroweak interactions, also encompassing strong nuclear interactions, and a possible unification with gravity.

In the late 19th and early 20th centuries, gravity and electromagnetism were known as fundamental forces of nature, so it is not surprising that distinguished physicists and mathematicians tried to unify these two interactions, following Maxwell's success with electricity and magnetism. For instance, Gustav Mie proposed that the electron was something like a "knot" (*Knotenstellen*) of the electromagnetic field in the Ether. For this purpose, it was necessary to generalize Maxwell's equations to permit such a structure to be stable.[1] As mentioned in Chap. 4, Mie tried to develop such a theory from a principle of least action applied to a Lagrangian function that depended only on quantities that are invariant under changes of reference frame, that is, their values do not change under a Lorentz transformation.[2] Of course, it was later found that the electron is much more than a knot of electromagnetic field.

David Hilbert, being a great mathematician, wanted to establish the foundations of mathematics with only a limited system of clear and intuitive axioms, from which all the fundamental theorems and equations could be rigorously deduced. His ambitious program for mathematics stumbled in Gödel's theorem, but his idea of axiomatizing

[1] Corry (1999), Corry, Renn and Stachel (1997).
[2] The electric and magnetic fields, E and B, admit two invariants, E^2-B^2 and the scalar product $E \cdot B$ (in Gaussian units), they keep the same value in any reference system. Mie also added terms involving the electromagnetic potential A_i.

physics led to an alternative derivation of the equations of general relativity (as we saw in Chap. 3). In his renowned article *The Foundations of Physics, Part One,*[3] he proposed two fundamental axioms for a theory of gravity and electromagnetism:

1) There is a "world function" (*Weltfunktion*) which contains only the metric tensor g_{ij} as potentials, together with their first and second derivatives, to which Hilbert, following Mie, added terms with the electromagnetic field and its potential vector A_i.
2) The *Weltfunktion* must be invariant under transformations from one reference system to another.

Such a *Weltfunktion* must also be the simplest possible, which is why Hilbert chose the Ricci scalar,[4] from where, with the help of Hamilton's principle of least action, he derived Einstein's equations of general relativity. Originally, Hilbert had also included a generalized electromagnetic field, as the one proposed by Mie, because he was looking for a unified theory of gravitation and electromagnetism. That terms, however, can be replaced with a true Lagrangian describing the presence of matter, and thus the full Einstein equations with matter follow directly.

16.2   Dirac and Mathematical Beauty

In a paper presented in 1939,[5] Dirac also reflected on the aesthetical aspects of mathematics, in addition, of course, to its usefulness. He pointed out that classical physics had the great merit of being based on a few relatively simple laws, but that simplicity could no longer be found in the two theories of modern physics, relativity and quantum mechanics. Nevertheless, these theories could be described as mathematically...beautiful! Of course, the concept of beauty in mathematics is as subjective as art. Poincaré had already pointed out that mathematics has an important aesthetic aspect, since its adepts[6] "find pleasure in it analogous to those afforded by painting and music. They admire the delicate harmony of the numbers and shapes...".

A good example of mathematical aesthetics is the Dirac equation that describes an electron (the formula appears on a memorial stone, in Westminster, at the feet of Isaac Newton). The beauty of this equation, as a physicist or a mathematician would say, lies in the fact that it condenses a profound physical truth in a very few symbols. Each of these symbols represent very complex mathematical concepts, but form a surprising synthesis in one equation. Something similar could be said of Euler's formula that condenses the most basic numbers of mathematics (π, e, 0, 1, i) into a short line.

---

[3] Hilbert (1915). *Die Grundlagen der Physik. Erste Mitteilung.*

[4] See Chap. 3.

[5] Dirac (1939), *The Relation between Mathematics and Physics.*

[6] Poincaré, *L'analyse et la physique,* in *La valeur de la science,* Poincaré (1970).

Having clarified the above, Dirac proposed that there are two parallel worlds, mathematics and real nature, closely related to each other. Thus, the true facts of the real world could be found searching for their equivalent concepts in the world of mathematics. "Possibly the two subjects will ultimately unify, every branch of pure mathematics then having its physical application, its importance in physics being proportional to its interest in mathematics".[7] In this process of research, mathematical beauty would be an important guide, along with experiments.

Continuing with his idea, Dirac also ventured into cosmology with an intriguing proposition. He noted that the ratio of electric to gravitational forces between an electron and a proton (e.g., a hydrogen atom) is a disproportionately large number, of the order of 10^{40}. Curiously, that number also corresponds, in order of magnitude, to the radius of the visible Universe measured in the typical units of the atomic world, the so-called classical radius of the electron.[8] Dirac speculated that this was not a coincidence, but rather an indication that the forces of gravity and electromagnetism had been equal in the "beginning" of the Universe, but that they have separated since then. If so, Newton's constant of gravity G would not be a constant, and its value should have diminished with time as $1/t$, where t is the age of the Universe.

Dirac's hypothesis aroused some interest as an alternative theory of cosmology not based on general relativity, but it is nowadays no longer taken seriously due to the great successes of the Big Bang theory. The extremely precise measurements that can be made today have not, so far, detected any variation of the constant of gravity.

16.3   Eddington and Numerology

A true "final theory" or "theory of everything" should explain with a very limited number of postulates the values of the natural constants: for instance, the charge and mass of the electron or the quarks... or even Planck's or Newton's constants, etc.? But what is it meant by such postulates? Would it be a theory based on purely mathematical grounds?

There were some attempts to elaborate such a theory in the first half of the twentieth century, the main exponent being the distinguished astrophysicist Arthur Eddington, whom we already met as one of the most important pioneers of general relativity. His ideas were published posthumously with the title *Fundamental Theory*,[9] but it was never taken seriously by the scientific community and was soon forgotten. Nevertheless, Eddington´s theory is a good example of what should be, in the most extreme case, a "theory of everything".

---

[7] Dirac, *op. cit.*

[8] The classical radius r_e of the electron is defined as the distance at which the potential energy of an electron bound to a proton, e^2/r_e, equals its energy in the form of mass mc^2 (e charge and m mass of the electron). It is $r_e = e^2/mc^2$, equivalent to about 2.8×10^{-15} m.

[9] Eddington (1947), *Fundamental Theory*.

For Eddington, the existence of pure numbers in Nature —that is, natural constants that do not depend on the system of units— was very significant. For instance, the ratio between electric and gravitational forces, or the relation between the charge e of the electron and hc (Planck's constant multiplied by the speed of light), which physicists call α and define as

$$\alpha = e^2/\hbar c$$

The latter is a dimensionless number, the *fine structure constant*[10] and represents the strength of electromagnetic forces at the atomic level. The first measurements of this constant in the early twentieth century established an approximate (inverse) value of $\alpha \approx 1/137.0$, which led to some speculations among physicists as to whether the inverse of α might be an exact integer.[11] (For informational purposes, the most recent measured value was $\alpha = 1/137.0359990\ldots$).

With rather obscure arguments, Eddington concluded that the integer 136 should be fundamental for a theory encompassing both the atomic world and cosmology. As the fine structure constant turned out to be $\alpha \approx 1/137.0$ when more precise measurements were performed, Eddington concluded that 137 would the fundamental number of Nature. In particular, he speculated, based on weird arguments, that the number $N = (3/2) \times 136 \times 2^{2 \times 136}$ should be, in some way, the exact (!) number of protons and electrons in the Universe. Moreover, in his own words: "But N has a more general significance as a fundamental constant which enters into many physical formulae; it determines the ratio of the electrical to the gravitational force between particles, the range and magnitude of the non-Coulombian forces in atomic nuclei, and the cosmical repulsion manifested in the recession of nebulae."[12]

Not surprisingly, Eddington's speculations were ignored by his colleagues, despite his well-deserved and unquestioned fame as an astrophysicist. We now know that Nature is much more complicated than what Eddington thought in his time, with a plethora of elementary particles and forces between them that are not only electromagnetic and gravitational. His attempt to build basic physics on mathematical models and speculations is an extreme position, but there have been less ambitious and more realistic attempts to discover physical reality with mathematical speculations. In fact, there are still distinguished mathematical physicists who have tried to follow similar lines, especially in their popular books aimed to impress lay public.[13]

---

[10] It is related to the splitting (fine structure) of the spectral lines of the hydrogen atom.

[11] See, for instance, Miller (2010).

[12] Eddington, *op cit*, appendix.

[13] See, for instance, the popular texts by S. W. Hawking and Roger Penrose.

16.4 Laws of Nature

All of the above leads to the question of what the origin of the laws of Nature is. Shortly after his eclipse expedition, Eddington wrote an article for the journal *Mind*,[14] in which he argued that the laws of Nature would be products of our mind and not properties of the material world. Eddington remarked that Einstein's equations in vacuum had a term that one sets equal to zero in the absence of matter. This means that what is interpreted as vacuum in this theory is the space that contains the gravitational field only (this is the case of the Schwarzschild and Kerr metrics), or what is the same, the curvature of space–time without matter.[15] But then, Eddington pointed out, it follows from Einstein's equations that "Einstein's law of gravitation is not a law of Nature but a definition—the definition of a vacuum". On the other hand, when the terms of these equations, related to matter and energy, they were not equal to zero... "the corresponding property of the world is perceived by us as a distribution of matter". Accordingly,

> Mind surveying the external world passes over unnoticed many of the differences of quality which from the mathematical standpoint are most elementary; it has developed no faculty for perceiving the quality measured by g_{ij}; but we have now arrived in our discussion at a quality which mind takes cognisance of and recognizes under the name of "emptiness".

Eddington's argument reminds us of the definition of "force" in classical mechanics. Newton's first law defines the absence of force as the state of rest or inertial motion, and the deviation from that state is attributed to a force. Similarly, the absence of matter and energy would manifest as the local presence of the gravitational field and nothing else. This is how something like vacuum would be conceived.

From the above arguments, it seems clear that Eddington was attempting to reconstruct physics from mathematical elaborations based on some a priori knowledge. As Helge Kragh pointed out,[16] Eddington's position was quite consistent with a scientific-philosophical trend that occurred in his time, mainly in the United Kingdom, and in which other distinguished mathematical physicists participated (Kragh, 2017):

> To put it briefly, there existed in Britain in the 1930s a fairly strong intellectual and scientific tradition that in general can be characterized as anti-empirical and pro-rationalist, although in some cases the rationalism was blended with heavy doses of idealism. According to scientists associated with this attempt to rethink the foundation of physical science, physics was inextricably linked to cosmology. In their vision of a future fundamental physics, pure thought counted more heavily than experiment and observation. The leading cosmo-physicists of the interwar period... were Eddington and E. Arthur Milne, but also Dirac, James Jeans and several other scientists held views of a roughly similar kind.

---

[14] Eddington (1920), *The Meaning of Matter and the Laws of Nature according to the Theory of Relativity*.
[15] Eddington, *op cit*.
[16] Kragh, (2017), *On Arthur Eddington's Theory of Everything*.

Arthur Milne (1896–1950) developed his own cosmological theory. James Jeans (1877–1946) is known mainly for his work on the formation and evolution of stars. Both were renowned English astrophysicists.

16.5 The Dream of Unification

Dirac's hypothesis that gravity varies with the evolution of the Universe or Eddington's deduction of the natural constants from numerical manipulations, are examples of the attempts by some theoretical physicists to unify the laws of the atomic world and the Universe.

It must be remembered that at the beginning of the twentieth century, only two fundamental forces of Nature were known, so that the attempts to find a unified theory included only these two interactions. Einstein himself dedicated his last years at Princeton to the search of such a unifying theory. Unfortunately, all the attempts were unsuccessful because Nature turned out to be much more complex at the atomic level. The only unification that has been achieved to date is that of electromagnetism with the weak nuclear force, for which it was necessary to take into account the existence of particles of the same "family" as the photon, the Z and W bosons, but with enormous masses that have not been explained until now.

Currently, the closest thing to a final (working!) theory is the Standard Model of elementary particles, which describes strong and electroweak interactions in a unified way. However, it is a model and not a theory since it includes twenty-four elementary particles whose masses and interactions are not explained but must be fixed "by hand".[17] It would also be necessary to include the elusive neutrinos in an extension of the model. And last but not least, the gravitational force, as was attempted (unsuccessfully) with superstring theory.

In any case, very few physicists at the present time still believe that the basic laws of Nature can be deduced, at least partially, from a refinement of mathematical formalisms and concepts.

---

[17] See, for instance: Weinberg (2011), Dreams of a Final Theory.

Chapter 17
Conclusions: Pythagoras' Dream

> *The fact that [the world of our sense experiences] is*
> *comprehensible is a miracle.*
> Albert Einstein (Einstein (1936), op. cit.)

Abstract Without numbering, the world would be incomprehensible. It must be assumed that numbers do not change arbitrarily. Hence the question arises whether Gödel's theorem can impose some restrictions to the elaboration of the laws of Nature.

As we have seen, many physicists and mathematicians consider that there is beauty in mathematics, just as there is in any art. The Dirac equation is undoubtedly beautiful because it describes an electron very concisely with an ingenious mathematical "trick", to the extent that, for some physicists, the electron is simply the solution of the Dirac equation. However, matters have become more problematic as Nature reveals to be complicated. In the subatomic world, the famous Standard Model is given in terms of an enormous Lagrangian containing more than twenty terms and a dozen parameters that must be deduced from experiments. Though undeniably correct, no physicist would pretend that the Model is beautiful.

The Lagrangian of the Standard Model represents the elementary particles, as the painting of a pipe represents a pipe... but is not a pipe!

17.1 Blue Tigers

Could we understand the world without mathematics? At some point, it is unavoidable to count and assign numbers to perceived objects... but always with the expectation that what is being counted will be preserved. Without the principle of permanence, numbers would not make any sense. Borges, in his short story *Blue Tigers*, relates the misadventures of an explorer who finds some mysterious blue stones—the tigers—whose properties led him to the brink of madness. The stones were impossible to

count, since their number increased or decreased unpredictably, contradicting all
mathematical reasoning.

> The same longing for order that created mathematics in the beginning made me seek an
> order in that aberration of mathematics that are the senseless stones that engender. In their
> unpredictable variations I wished to find a law.[1]

But it was useless, there was no law.

What if such or similar stones existed in the real world, with the same properties?
They would violate the principle of permanence, so crucial to Kant, and mathematics
would be of no avail if atoms appeared and disappeared without law. However, due
to the principle of sufficient reason, so dear to Leibniz and Schopenhauer, scientists
would not accept the impossibility of counting, but would search for the reason why
these blue tigers appear and disappear, always with the conviction that there is a
hidden law that regulates their behavior, a law that must be revealed. Even in the
world of atoms, quantum particles do not reject the calculation of probabilities.

But in Borges's story, the protagonist tried various experiments with the stones,
without any success. "Mathematics... has its beginning and now its end in these
stones. If Pythagoras had dealt with them...".

Finally, the protagonist managed to get rid of the stones by handing them to a
mysterious beggar, but who left him a "terrible alm" in return: "You stay with the
days and the nights, with sanity, with habits, with the world".

17.2   The Realm of Numbers

The famous theorem of mathematical logic that we owe to Gödel stipulates that, no
matter which system of axioms is chosen, there are inevitably propositions that can
neither be proved nor disproved.

Gödel's original article, published in 1931, is extremely technical and only accessible to experts on the field,[2] but in essence, Gödel's procedure was to establish
a one-to-one relationship between any mathematical proposition and the integer
numbers. For this purpose, he noted that any text, including mathematical proofs,
can be represented by integers. For instance, a number can be associated with each
letter of the alphabet (say, from 1 to 26), so the content of this book would be a
certain number, obviously of enormous length, but an integer number nevertheless.

Now, any proof in mathematics starts from a set of axioms—a succession of
symbols—, which are transformed according to certain rules, step by step, until
reaching the result to be proved —another succession of symbols—. Thus, Gödel
found a particular way to associate a unique number to any general mathematical
proposition, which would be its "Gödel number", unique in the sense that the proposition from which it was produced can be recovered. Taking advantage of the properties

---

[1] J. L. Borges, *Tigres azules* (my translation).

[2] For a more accessible version, see: Nagel and Newman, (2008), *Gödel's proof*.

of prime numbers,[3] the next step was to prove that there exist propositions whose Gödel numbers cannot lead, by means of any established transformation rules, to any Gödel number that could correspond to its proof.[4]

Does Gödel's theorem have any relevance for the description of the material world? In this regard, Nagel and Newman warned their readers against jumping to rash conclusions about the laws of Nature[5]:

> The discovery that there are arithmetical truths which cannot be demonstrated formally does not mean that there are truths which are forever incapable of becoming known... It does mean that the resources of the human intellect have not been, and cannot be, fully formalized, and that new principles of demonstration forever await invention and discovery.

Thus, for instance, if new and unexpected experimental results were found that contradict the Standard Model, a more general Lagrangian function would have to be constructed. But it would always be a mathematical representation of the world, based on numbers and mathematical symbols and transformation rules between those symbols.

What Gödel's theorem shows is that our reasoning is in direct correspondence with integer numbers. Therefore, reason cannot be free of the rules of arithmetic. As Borges said, we remain "with sanity, with habits, with the world". And with everything that can be counted and represented by numbers and, in a more elaborated way, by mathematics... if the (a priori?) principles of permanence and sufficient reason are valid. Fortunately for our reason, the world can be represented by mathematics because there are no blue tigers... or so we want to think.

---

[3] Prime numbers (2, 3, 5, 7, 11, 13, 17, etc.) are those that are not the products of smaller integers (other than 1). A fundamental theorem is that any non-prime number can be represented by the unique product of prime numbers (for instance, $165 = 3 \times 5 \times 11$).

[4] Of course, this is a toy version of Gödel's theorem. For a more serious exposition, see Nagel and Newman, *op. cit.*

[5] Nagel and Newman, *op. cit.*

Bibliography

Aspect A et al (1982) Experimental realization of the Einstein-Podolsky-Rosen-Bohm *Gedanken-experiment*: A new realization of Bell´s inequalities. Phys Rev Lett 49: 91–94

Balaguer M (2001) Platonism and anti-Platonism in Mathematics. Oxford Univ Press, Oxford

Bell J (1987) Speakable and Unspeakable in Quantum Mechanics. Cambridge Univ Press, Cambridge

Blagojevic M, Hehl F W (eds) (2013) Gauge Theories of Gravitation: A Reader with Commentaries. Imperial College Press, London

Bohr N (1987) Quantum Physics and Philosophy: Causality and Complementarity. In: The Philosophical Writings of Niels Bohr, Vol III. Ox Bow Press, Woodbridge, Connecticut

Bouwmester D, Ekert AK, Zeilinger A (2000) The Physics of Quantum Information. Springer, Heidelberg

Calinger RS (2019) Leonhard Euler: Mathematical Genius in the Enlightenment. Princeton University Press, Princeton

Calvino I (1995) Ti con Zero. Mondadori, Italy

Carnap R (1928) Der logische Aufbau der Welt. Weltkreis-Verlag, Berlin-Schlachtensee. English edition: Carnap R (1967) The Logical Structure of the World (trans: George RA). Univ California Press

Chandrasekhar S (1995) Newton's Principia for the Common Reader. Oxford University Press, Oxford

Clauser JF, Shimony A (1978) Bell's theorem. Experimental tests and implications. Rep Prog Phys 41:1881–1928

Corry L, Renn J, Stachel J (1997) Belated Decision in the Hilbert-Einstein Priority Dispute. Science 278:1270-1273

Corry L (1999) From Mie's Electromagnetic Theory of Matter to Hilbert's Unified Foundations of Physics. Studies Hist Phil Mod Phys 30:159–183.

Crowe MJ (1967) A History of Vector Analysis: The Evolution of the Idea of a Vectorial System. Dover, New York

Darrigol O (2012) A History of Optics. Oxford

Dirac PAM (1939) The Relation between mathematics and physics. Proc Roy Soc (Edinburgh) 59 Part II: 122–129

Dugas R (1950) Histoire de la Mécanique. Dunod, Paris. English edition: Dugas R (1988) History of Mechanics (trans: Maddox JR). Dover, New York

Duhem P (1905) L'Évolution de la Mécanique. Hermann, Paris. English edition: Duhem P (1980) The Evolution of Mechanics (trans: Cole JM). Springer, Heidelberg

S. Hacyan, *The Mathematical Representation of Physical Reality*, The Frontiers Collection, https://doi.org/10.1007/978-3-031-21254-3

Duhem P (1906) La Théorie Physique: son Object, et sa Structure. Chevalier & Rivière, Paris. English edition: Duhem P (1991) The Aim and Structure of Physical Theory (trans: Wiener PP). Princeton U Press, Princeton

Dunham W (1999) Euler: The Master of Us All. American Mathematical Soc

Eddington AS (1920) The meaning of matter and the laws of nature according to the theory of relativity. Mind 29 (114):145–158

Eddington AS (1923) The Mathematical Theory of Relativity. Cambridge Univ Press, Cambridge

Eddington AS (1949) Fundamental Theory. Cambridge Univ. Press, Cambridge

Einstein A (1934) Mein Weltbild. Querido Vlg, Amsterdam. English edition: Einstein A (1935) The World as I See It (trans: Harris A). Bodley Head, London

Einstein A (1936) Physics and Reality. J Franklin Inst 221(3):349-382

Einstein A (1954) Preface. In: Jammer M (1954) Concepts of space. Dover, New York

Eisenhart LP (1959) Riemannian Geometry. Princeton Univ Press. Princeton

Euler L (1736) Mechanica. English translation Bruce I. http://www.17centurymaths.com/contents/mechanica1.html

Feferman S et al (1995). Kurt Gödel: Collected works, Vol. III. Oxford Univ Press, Oxford

Feyerabend P (1975) Against Method. New Left Books (verso), New York, London

Feyerabend P (1999) Conquest of Abundance. A Tale of Abstraction versus the Richness of Being. Univ Chicago Press, Chicago

Feynman R, Leighton R, Sands M (1968) The Feynman Lectures on Physics Vol. III. Addison & Wesley, California

Feynman R, Leighton R, Sands M (2011) Six Easy Pieces: Essentials of physics explained by its most brilliant teacher. Basic bookx, New York

Galilei G (1623) Il Saggiatore. Galileo G (1960) The Assayer (trans: Stillman D & O'Malley CD in: The Controversy on the Comets of 1618. Univ Pennsylvania Press, Pennsylvania

Gödel K (1949) A remark about the relationship between relativity theory and idealistic philosophy. In: Schilpp P (ed) Albert Einstein: Philosopher Scientist. Evanston, Illinoi

Guth AH (2000). Inflation and eternal inflation. Physics Reports 333–334:555–574

Hacyan S (2004) Física y metafísica del espacio y el tiempo. Fondo de Cultura Económica, Mexico (In Spanish)

Hacyan S (2006) On the Transcendental Ideality of Space and Time in Modern Physics. Kant-Studien 97:382–395

Hacyan S (2009) Geometry as an object of experience: The missed debate between Poincaré and Einstein. European J Physics 30:337–343

Hacyan S (2015) Decoherence and the paradox of the observed observer in quantum mechanics. Physica Scripta 90:074001 (4pp)

Heering P (1992) On Coulomb's inverse square law. American J Phys 60:988–993

Heisenberg W (1958) Physics and Philosophy. Harper, New York

Heisenberg W (1972) Quantum mechanics and a talk with Einstein. In: Physics and beyond. Encounters and conversations (trans Pomerans AJ). Harper Torchbooks, New York

Heisenberg W (1990) Across the Frontiers (trans: Heath P). Ox Bow Press, Connecticut

Helmholtz H (1995). On the origin and significance of geometrical axioms. In: Cahan D (ed) Science and Culture. Univ Chicago Press, Chicago

Hesse MB (1962) Forces and Fields. Greenwood, Connecticut

Hilbert D (1915) Die Grundlagen der Physik. (Erste Mitteilung) Nachrichten von der Königlichen Gesellschaft der Wissenschaften zu Göttingen. Mathematisch-physikalische Klasse: 395–407

Hilbert D (2009) In: David Hilbert's Lectures on the Foundations of Mathematics and Physics, 1891–1933. Sauer T, Majer U (eds). Springer-Verlag, Heidelberg

Husserl E (1954) The Crisis of European Sciences and Transcendental Phenomenology (trans: Carr D). Northwestern Univ. Press, Illinois

Jammer M (1961) Concepts of Mass in Classical and Modern Physics. Harvard Univ Press, Harvard

Jammer M (1993) Concepts of space. The history of theories of space in physics. Dover, New York

Kant I (1929) Critique of Pure Reason (trans: Smith NK). MacMillan & Co, Boston/New York

Kant I (2009) Metaphysical Principles of the Science of Nature, preface. (trans ed: Bennett J) https://www.earlymoderntexts.com/assets/pdfs/kant1786_2.pdf

Kant I (2012) Thoughts on the True Estimation of Living Forces. Watkins E (ed). Cambridge, Cambridge https://doi.org/10.1017/CBO9781139014380

Kennnefick D (2007) Traveling at the Speed of Thought: Einstein and the Quest for Gravitational Waves. Princeton Univ Press, Princeton

Koestler A (1959) The Sleepwalkers: A History of Man's Changing Vision of the Universe. Hutchinson, UK

Kragh H (2017) On Arthur Eddington's Theory of Everything. In: Durham I, Rickles D (eds) Information and interaction: Eddington, Wheeler, and the limits of knowledge. Springer, Heidelberg

Landau LD, Lifshitz EM (1975) The Classical Theory of Fields. Elsevier, Amsterdam

Leibniz GW (1956) Philosophical Papers and Letters, Loemker LE (ed). Reidel, Dordrecht

Leibniz GW (1678–1989) A brief demonstration of a notable error of Descartes and others concerning a natural law. In: Loemker LE (ed) Philosophical Papers and Letters. The New Syntheses Historical Library, vol 2. Springer, Dordrecht https://doi.org/10.1007/978-94-010-1426-7_35

Mach E (1883) Die Mechanik in ihrer Entwicklung: Historisch-Kristisch Dargestellt. Akademie, Berlin. English edition: Mach E (1942) The science of mechanics: A critical and historical account of its development (trans: McCormack TJ). Open Court, Illinois

Maxwell JC (1873) A Treatise on Electricity and Magnetism. Oxford Univ Press, Oxford. Facsimile edition (1998). Oxford Classics Texts in the Physical Sciences

Maxwell JC (1952) Matter and Motion. Dover, New York

Mehra J (1974) Einstein, Hilbert and the Theory of Gravitation. Reidel, Dorchtrech/Boston

Mie G (1914) Königliche Gesellschaft der Wissenschaften zu Göttingen. Nachrichten: 2336

Miller AI (2010) 137: Jung, Pauli, and the Pursuit of a Scientific Obsession. Norton, New York

Nagel E, Newman JR (2008) Gödel's Proof (Revised). Routledge & Kegan, London

Pauli W (1949) The influence of archetypal ideas on the scientific theories of Kepler. In: Enz CP, von Meyenn K (eds) Wolfgang Pauli. Writings on Physics and Philosophy (trans: Schlapp R). Springer-Verlag, Berlin/London

Poincaré H (1968) La Science et l'Hypothèse. Flammarion, Paris. English edition: Poincaré H (2018) Experiment and Geometry. In: Science and Hypothesis (trans: Frappier M, Smith A, Stump DJ). Dover, New York

Poincaré H (1970) La valeur de la science. Flammarion, Paris. English translation: Poincaré H (2007) The Value of Science (trans: Halstead GB). Cosimo, New York

Reichenbach H (1958) The Philosophy of Space and Time. Dover, New York

Renn J, Stachel J (2007) Hilbert's foundation of physics: From a theory of everything to a constituent of general relativity. In: Renn J, Schemmel M (eds) The Genesis of General Relativity. Springer, Dordrech, p 857–974

Renn J, Schemmel M (eds) (2007) The Genesis of General Relativity. Volume 4: Gravitation in the Twilight of Classical Physics, The Promise of Mathematics. Springer, Dordrecht

Ribe N, Steinle F (2002) Exploratory Experimentation: Goethe, Land, and Color Theory. Physics Today 55(7):43–49

Roche J (2003) What is Potential Energy? Eur J Phys 24:185–196

Rutz F (1993) A Finsler generalization of Einstein's vacuum field equations. Gen Rel Grav 25:1139–1158

Schilpp PA (ed) (1949) Albert Einstein: Philosopher–Scientist. Cambridge Univ Press, Cambridge

Schopenhauer A (1966) English edition: The World as Will and Representation, Vol. I and II (trans: Payne EF). Dover, New York

Schopenhauer A (1974) English edition: On the Fourfold Root of the Principle of Sufficient Reason (trans: Payne EF). Open Court, Illinois

Smeenk C, Martin C (2007) Mie's theories of matter and gravitation. In: Janssen M, Norton JD, Renn J, Sauer T, Stachel J. (eds) The Genesis of General Relativity, Boston Studies in the Philosophy of Science, vol 250. Springer, Dordrecht

Suisky D (2009) Euler as a Physicist. Springer, Heidelberg

Tolman R (1934) Relativity, Thermodynamics and Cosmology. Clarendon, Oxford

Tonomura A, Endo J, Matsuda T, Kawasaki T, Ezawa H (1989) Demonstration of single-electron buildup of an interference pattern. American J Phys 57:117–120

Utiyama R (1956) Invariant theoretical interpretation of interaction. Phys Rev 161:597–607

Weinberg S (2008) Cosmology (Appendix F p 543). Oxford Univ Press, Oxford

Weinberg S (2011) Dreams of a Final Theory: The Scientist's Search for the Ultimate Laws of Nature. Random House, New York

Westfall RS (1980) Never at Rest. A Biography of Isaac Newton. Cambridge Univ Press, Cambridge

Wheeler GF, Crummett WP (1987) The vibrating string controversy. American J Phys 55:33–37

Whittaker ET (1919) A History of the Theories of Aether: From the Age of Descartes to the Close of the Nineteenth Century. Reprinted by Kessinger Publishing, Montana

Wigner EP (1960) The unreasonable effectiveness of mathematics in the natural sciences. Comm Pure Applied Math 13:1–14

Wigner EP (1961) Remarks on the mind-body question. In: Good IJ (ed) The Scientist Speculates. Heinemann, London, p 284–302

Wittgenstein L (1956) Remarks on the Foundations of Mathematics. Blackwell, New Jersey

Woit P (2007) Not even wrong: The Failure of String Theory and the Search for Unity in Physical Law. Basic Books, New Yorks

Yang CN, Mills RL (1954) Conservation of isotopic spin and isotopic gauge invariance. Phys Rev 96:191–195

Youschkevitch AP (1976) The concept of function up to the middle of the 19th century. Arch. Hist. Exact Sci. 16: 37–85 https://doi.org/10.1007/BF00348305

Index

Printed in the United States
by Baker & Taylor Publisher Services

Printed in the United States
by Baker & Taylor Publisher Services